Ultimate limit-state design of concrete structures: a new approach

M.D. Kotsovos and M.N. Pavlović

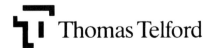

Published by Thomas Telford Ltd, 1 Heron Quay, London E14 4JD
URL: http://www.t-telford.co.uk

Distributors for Thomas Telford books are
USA: ASCE, 1801 Alexander Bell Drive, Reston, VA 20191-4400
Japan: Maruzen Co. Ltd, Book Department, 3–10 Nihonbashi 2-chome, Chuo-ku, Tokyo 103
Australia: DA Books and Journals, 648 Whitehorse Road, Mitcham 3132, Victoria

First published 1999

A catalogue record for this book is available from the British Library

ISBN: 0 7277 2665 X

© M. D. Kotsovos and M. N. Pavlović, 1999

All rights, including translation reserved. Except for fair copying, no part of this publication may be reproduced, stored in a retrieval system or transmitted in any form or by any means, electronic, mechanical, photocopying or otherwise, without the prior written permission of the Books Publisher, Thomas Telford Ltd, 1 Heron Quay, London E14 4JD.

This book is published on the understanding that the author is/authors are solely responsible for the statements made and opinions expressed in it and that its publication does not necessarily imply that such statements and/or opinions are or reflect the views or opinions of the publishers.

Typeset by MHL Typesetting, Coventry.
Printed and bound in Great Britain by Redwood Books, Trowbridge, Wiltshire.

Contents

Preface   vii

1. An appraisal of the validity of predictions of current design methods   1

1.1. Introduction   1

1.2. Structural walls under transverse loading   2

1.3. Simply-supported reinforced concrete T-beams   5

1.4. Structural wall under combined vertical and horizontal loading   9

1.5. Column with additional reinforcement against seismic action   12
     1.5.1. Design details   12
     1.5.2. Experimental behaviour   14
     1.5.3. Causes of failure   16

1.6. Simply-supported beams with one overhang   18
     1.6.1. Design details   18
     1.6.2. Experimental behaviour   19
     1.6.3. Causes of failure   23

1.7. Failure of structural concrete under seismic load   24

1.8. Conclusions   28

1.9. References   29

2. Reappraisal of current methods for structural concrete design   31

2.1. Introduction   31

2.2. Current design methods   31
     2.2.1. Physical model   31
     2.2.2. Beam action   34
     2.2.3. Truss action   36

2.3. Reappraisal of the current approach for assessing flexural capacity 38
    2.3.1. Concrete behaviour 39
        2.3.1.1. Stress–strain curves 39
        2.3.1.2. Cracking 41
        2.3.1.3. Effect of small transverse stresses on strength and deformation 43
    2.3.2. Failure mechanism of the compressive zone 45
        2.3.2.1. A fundamental explanation of failure initiation based on triaxial material behaviour 45
        2.3.2.2. Triaxiality and failure initiation by macrocracking: some experimental and analytical evidence 49

2.4. Reappraisal of the current approach for assessing shear capacity 62
    2.4.1. Validity of concepts underlying shear design 62
        2.4.1.1. Shear capacity of critical cross-section 63
        2.4.1.2. Aggregate interlock 66
        2.4.1.3. Dowel action 66
        2.4.1.4. Truss analogy 66
    2.4.2. Contribution of compressive zone to shear capacity 67
    2.4.3. Shear-failure mechanism 69
    2.4.4. Contribution of transverse reinforcement to shear capacity 71

2.5. Conclusions 74

2.6. References 75

3. The concept of the compressive-force path 77

3.1. Introduction 77

3.2. Proposed function of simply-supported beams 77
    3.2.1. Physical state of beam 77
    3.2.2. Load transfer to supports 79
    3.2.3. Effect of cracking on internal actions 82
    3.2.4. Contribution of uncracked and cracked concrete to the beam's load-carrying capacity 84
    3.2.5. Causes of failure 85

3.3. Validity of the proposed structural functioning of simply-supported beams 88

3.4. Conclusions	94
3.5. References	95
4. Design methodology	**96**
4.1. Introduction	96
4.2. Simply-supported reinforced concrete beam	96
4.2.1. Physical model 96	
4.2.2. Failure criterion 98	
4.2.3. Validity of failure criteria 102	
4.2.4. Assessment of longitudinal reinforcement 108	
4.2.5. Assessment of transverse reinforcement 110	
4.2.6. Design procedure 115	
4.2.7. Design examples 116	
(a) Beam of type II behaviour 117	
(b) Beam of type III behaviour 120	
(c) Beam of type IV behaviour 123	
4.3. Simply-supported prestressed concrete beam	124
4.3.1. Physical model 124	
4.3.2. Failure criteria 127	
4.3.3. Assessment of reinforcement 127	
4.3.4. Procedure for checking shear capacity 127	
4.3.5. Example of shear-capacity checking 129	
4.4. Skeletal structural forms with beam-like elements	134
4.4.1. Physical models 134	
4.4.2. Design procedure 137	
4.4.3. Design examples 137	
(a) Simply-supported beam with over-hang 138	
(b) Cantilever 141	
(c) Continuous beam 144	
(d) Portal frame with fixed ends 149	
4.5. The failure of an offshore platform	154
4.5.1. Background 154	
4.5.2. A simple structural evaluation 157	
4.5.3. Strength evaluation of test specimens 160	
4.5.4. Concluding remarks 163	
4.6. References	163

Preface

The adoption of the limit-state philosophy as the basis of current codes of practice for the design of concrete structures expresses the conviction that this philosophy is capable of leading to safer and more economical design solutions. After all, designing a structural concrete member to its ultimate limit state requires the assessment of the load-carrying capacity of the member and this provides a clearer indication of the margin of safety against collapse. At the same time, the high internal stresses which develop at the ultimate limit state result in a reduction of both the size of the member cross-section and the amount of reinforcement required to sustain internal actions. (Admittedly, the latter economy and, of course, safety itself are dependent on the actual factor of safety adopted; nevertheless, the more accurate estimate of the true failure load provides an opportunity to reduce the uncertainties reflected in the factor of safety in comparison with, say, elastic design calculations.)

In contrast to the above expectations for more efficient design solutions, recent attempts to investigate experimentally whether or not the aims of limit-state philosophy for safety and economy are indeed achieved by current codes of practice have yielded conflicting results. Experimental evidence has been published that describes the behaviour of a wide range of structural concrete members (such as, for example, beams, columns, structural walls, slabs, etc.) for which current methods for assessing structural performance yield predictions exhibiting excessive deviations from the true behaviour as established by experiment. In fact, in certain cases the predictions underestimate considerably the capabilities of a structure or member — indicating that there is still a long way to go in order to improve the economy of current design methods — while in other cases the predictions are clearly unsafe as they overestimate the ability of a structure or member to perform in a prescribed manner; and this provides an even more potent pointer to the fact that a rational and unified design methodology is still lacking for structural concrete.

In the authors' book, *Structural Concrete*, published in 1995, the investigation of the causes of the above deviations led to the conclusion that the conflicting predictions are due to the inadequacy of the theoretical basis of the design methods which are used to implement the limit-state philosophy in practical design, rather than to the unrealistic nature of the aims of the design philosophy as such. In fact, it was repeatedly proven in the book that the fundamental assumptions of the design methods which describe the behaviour of concrete at both the material and the structure levels were adopted as a result of misinterpretation of the available experimental information and/or use of concepts which, while working well for other materials (e.g. steel) or regimes (e.g. elastic behaviour), are not necessarily always suitable to concrete structures under ultimate-load conditions. Therefore, it becomes clear that the theoretical basis of current design methods requires an extensive revision if the methods are consistently to yield realistic predictions as a result of a rational and unified design approach.

Such a revision has been the subject of comprehensive research work carried out by the authors over the past decade. This was done concurrently at two levels. One of these — the higher level — was based on formal finite-element (FE) modelling of structural concrete with realistic material properties and behaviour as its cornerstone: most of the ensuing results are contained in the aforementioned book. At the second — the lower — level, an attempt was made to reproduce the essential results of complex numerical computations by means of much simpler calculations which would require no more effort than is the case with current code provisions. The latter approach was deemed necessary because, although the authors' FE model has proved useful as a consultancy tool for the design, redesign, assessment and even upgrading of reinforced concrete structures, the fact remains that most design offices still rely on simplified calculation methods which, if not quite 'back-of-the-envelope' stuff, are quick, practically hand-based (or easily programmable), provide (or claim to provide) a physical feel for the problem, and, of course, conform to the simple methodology of code regulations. The alternative methodology at this level, which stems from the authors' work and is the subject of the present book, and which may provide the basis for a new, improved design approach for the implementation of the limit-state philosophy into the practical design of concrete structures, involves, on the one hand, the identification of the regions of a structural member or structure through which the external load is transmitted from its point of application to the supports, and, on the other hand, the strengthening of these regions so as to impart to the member or structure desired values of load-carrying capacity and ductility. As most of the above regions enclose the trajectories of internal compressive actions, the new methodology has been termed the 'compressive-force path' (CFP) method. In contrast to the methods implemented in current codes of practice, the proposed methodology is fully compatible with the behaviour of concrete (as described by valid experimental information) at both the material and the structure levels.

Although the CFP method might appear, at first sight, to be a rather unorthodox way of designing structural concrete, it is easy,

PREFACE

with hindsight, to see that it conforms largely to the classical design of masonry structures by Greek and Roman engineers. These tended to rely greatly on arch action — later expressed (and extended) through the Byzantine dome and the Gothic vaulting. Now, such a mechanism of load transfer may seem largely irrelevant for a beam exhibiting an elastic response. However, for a cracked reinforced concrete girder close to failure the parallel with an arch-and-tie system reveals striking similarities between the time-honoured concept of a compressive arch and the newly-proposed CFP method.

The aim of the present book, therefore, is to introduce designers to the 'compressive-force path' method. Such an introduction not only includes the description of its underlying theoretical concepts and their application in practice but, also, presents the causes which led to the need for a new design methodology for the implementation of the limit-state philosophy into practical structural concrete design together with evidence — both experimental and analytical — supporting its validity. The book is divided into four chapters. Chapter 1 presents characteristic cases of structural concrete members, mostly designed in compliance with current code provisions, for which the behaviour predicted by the methods implemented in such code provisions deviates excessively from that established by experiment. In Chapter 2, the results obtained from an investigation of the fundamental causes which led to the excessive deviations discussed in the preceding chapter are described. These results from Chapter 2 are summarised in Chapter 3 so as to then form the theoretical basis of the proposed design methodology. The latter is fully described in Chapter 4, together with examples of its application in practical structural concrete design.

The authors are fully aware, of course, that code tenets cannot be ignored by the majority of designers, not only because of legal implications but, more positively, because many guidelines (especially those related to flexural failure) have been shown to provide sensible predictions for a wide range of reinforced concrete problems. Nevertheless, there are clearly problems (mainly those where 'shear' failures occur) for which code guidelines are less successful in their predictions, and such difficulties need to be addressed. The present book is intended to address such problems. Ultimately, however, it is up to the experienced engineer, as well as the young graduate or student well acquainted with present-day code rules, to decide whether or not ideas contained in this book do, in fact, provide a rational alternative to the design of structural concrete members.

The authors wish to express their gratitude to Dr Jan Bobrowski (of Jan Bobrowski & Partners, and Visiting Professor at Imperial College) for the many discussions they had with him on various aspects of structural concrete design: their research work has been strongly influenced by Jan's design philosophy and achievements, and this is certainly reflected in the method proposed in the book. The excellent collaboration established with Thomas Telford's editorial team during the preparation of the earlier book *Structural Concrete* has continued with the present work, its production proceeding in a relaxed and almost effortless manner under Linda Schabedly's authoritative editorial direction and the precise artwork of Lynne Darnell.

1. An appraisal of the validity of predictions of current design methods

1.1. Introduction   The design of a structural concrete member for ultimate strength requires the availability of methodologies capable of yielding realistic predictions of the member's ultimate characteristics, such as, for example, flexural capacity, shear capacity, ductility, etc. The efficiency of the resulting design solutions, therefore, is clearly dependent, to a large extent, on the ability of the methodology used to assess accurately these characteristics.

Current methodologies commonly employed for assessing the ultimate characteristics of a structural concrete member involve the use of analytical formulations which express the strength characteristics as a function of the member geometry and dimensions, as well as the mechanical properties of the materials from which the member is made. An important feature of these analytical formulations is the inclusion in them of empirical parameters, the evaluation of which is essential for the calibration of the formulations, such calibrations being achieved by using experimental data on the strength and deformational characteristics of the member. It is evident, therefore, that the very need to include these empirical parameters implies that the analytical formulations yield predictions which deviate from the corresponding values established by experiment.

Deviations up to approximately 10% are generally considered as natural since they are usually due to the scatter of the experimental data used for the calibration of the semi-empirical formulations relevant to the design method employed. On the other hand, deviations between approximately 10% and 20% are usually attributed to the lack (or deficiency) of experimental data sufficient to secure a conclusive calibration of the semi-empirical formulations. In the latter case, the acquisition of additional data may improve the calibration, thus leading to a reduction of the deviation of the predicted values of the strength characteristics from the true ones to its natural level which, as pointed out above, is of the order of 10%.

Deviations larger than 20% should be attributed to the lack of a sound theory underlying the derivation of the semi-empirical analytical formulations. In such a case, a reappraisal of the underlying theory is essential before an attempt is made to improve the prediction by acquiring additional experimental data for the more accurate calibration of the analytical formulations used.

To this end, the objective of the present chapter is the presentation of typical cases of structural concrete members, in most cases designed in compliance with current code provisions, the behaviour of which (as established by experiment) exhibits considerable deviation from the analytical predictions. For each case presented, attention is focused on identifying the particular formulations which cause the deviation of the predicted response from the measured response, with the aim of establishing the extent of the revision that current design methods require. The investigation of the fundamental causes of the inability of the above formulations to yield realistic predictions of the behavioural characteristics of structural concrete members forms the subject of the next chapter.

1.2. Structural walls under transverse loading

The experimental data, on which the discussion in the present section is based, are fully described in references 1.1 and 1.2. Therefore, in what follows, this experimental information will be presented only in a concise manner, as the aim is to highlight the most important features of the structural behaviour of the tested walls.

Figure 1.1 shows the geometric characteristics and dimensions of the walls together with the arrangement and diameter of the steel bars reinforcing a reinforced concrete (RC) structural wall (SWA) connected to upper and lower beams. (Throughout the book, all dimensions are in mm, unless stated otherwise.) The wall was clamped to the laboratory floor through the lower beam so as to form a fully-fixed connection, while the load was applied through

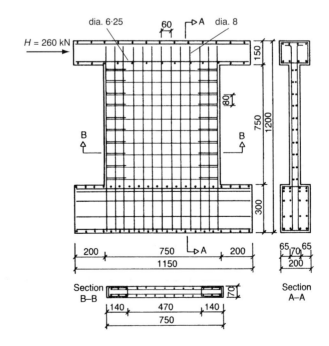

Fig. 1.1. Details of the structural walls designed to the codes[1.1,1.2]

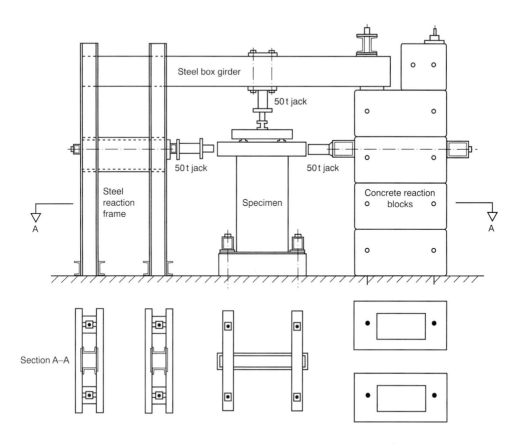

Fig. 1.2. Experimental set-up for the structural walls[1.1]

the upper beam as indicated in the experimental set-up depicted in Fig. 1.2. Although the wall behaviour was established under various combinations of vertical and horizontal loads, only the response under horizontal loading is discussed in the following.

The yield and ultimate stresses of the 8 mm diameter bars were 470 MPa and 565 MPa respectively, while for the 6 mm diameter bars the values of the above characteristics were 520 MPa and 610 MPa. The uniaxial cylinder compressive strength of concrete was 37 MPa.

By using the above material properties and geometric characteristics, the methods recommended by current codes yield values of the flexural and shear capacities equal to at least 200 kNm and 340 kN respectively, corresponding to values of the external load equal to 240 kN and 340 kN. (As the various codes share essentially the same premises, the actual load values obtained do not differ greatly from each other; for example, the present calculations encompassed the use of the following codes: BS 8110,[1.3] ACI 318-83,[1.4] Canadian,[1.5] CEB–FIP Model Code,[1.6] Greek.[1.7] Clearly, the wall's load-carrying capacity is the smaller of the above two values of the applied load, and, if

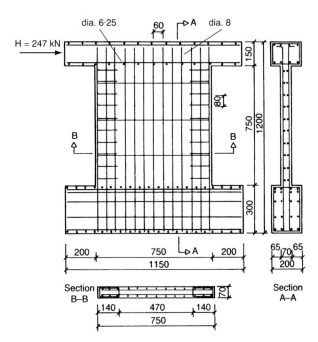

Fig. 1.3. Design details of the structural wall with horizontal reinforcement 30% of that specified by the codes

this is exceeded, flexural failure is expected to occur at the base of the wall, since the member's shear capacity is not expected to be exhausted for values of the applied load smaller than 340 kN.

In fact, by using the experimental set-up of Fig. 1.2 for subjecting the structural wall to a horizontal load (H) monotonically increasing to failure, it was found that the wall sustained $H = 260$ kN. This correlates closely with the value predicted by codes, the latter underestimating the experimentally established value by approximately 8%.

The structural wall SWB in Fig. 1.3 differs from wall SWA in Fig. 1.1 only in the number and spacing of the horizontal bars of the reinforcement. The bar spacing is 240 mm (i.e. three times the bar spacing in wall SWA) and, therefore, the total amount of the horizontal reinforcement is approximately 30% of that for wall SWA. As a result, in accordance with current code provisions, the contribution of the horizontal web reinforcement to shear capacity is reduced by about 67%, so that the predicted external load corresponding to shear capacity is now 155 kN. In fact, this value must represent the load-carrying capacity of wall SWB, since the load causing flexural failure remains the same as that of wall SWA (i.e. 240 kN).

Yet, in contrast with the case of wall SWA, the above prediction is not verified experimentally. In a repetition of the experiment carried out using wall SWA, wall SWB was found to fail again in 'flexure' and not, as now expected, in 'shear'. The load-carrying capacity was recorded as $H = 247$ kN, a value about 60% larger than that predicted through the usual design calculations. It would appear, therefore, that for the case of wall SWB, the methods recommended by current codes are incapable of predicting not only load-carrying capacity, but also the mode of failure.

If the predicted value of the shear capacity is ignored, the above results show that the methods currently used for the calculation of the flexural capacity yield realistic predictions. Such predictions were found to deviate from the experimental values by approximately 8%, which cannot be considered as indicative of the need for a reappraisal of the method used to assess flexural capacity. However, the need for a reappraisal of the methods currently used for assessing shear capacity appears to be mandatory. Although flexural failure of the walls did not allow the accurate assessment of the deviation of the predicted from experimental value of the shear capacity, it is evident that this deviation may exceed 60%, which is the deviation of the predicted shear capacity from the experimentally established flexural-failure value for wall SWB.

The finding that the transverse reinforcement required to prevent 'shear' failure is significantly smaller than that specified by present code provisions is indicative of the fact that current codes overestimate the contribution of such reinforcement at the expense of the contribution of other strength reserves which may exist in concrete or in the interaction between concrete and steel. The identification of such strength reserves and the assessment of their contribution to the total shear capacity may lead to a significant reduction of the amount of transverse reinforcement: this outcome could be associated not only with more efficient design solutions, but also, with an increase of the margin of safety against failure, as will be shown in the case studies that follow.

1.3. Simply-supported reinforced concrete T-beams

A more accurate indication of the size of the deviation of the value of shear capacity established by experiment from that predicted by the methods adopted by current codes resulted from a test programme concerned with an investigation of the behaviour of simply-supported RC T-beams (designed to the 'compressive-force path' (CFP) method described in detail in Chapter 4) under two-point and four-point loading, symmetrical with respect to the beam cross-section.[1.8] As for the case of the structural walls discussed in the preceding section, the full experimental data on the T-beams and their behaviour may be found in the literature.[1.8] In what follows, only a concise description of the experimental details is provided, the emphasis focusing on the main results.

As indicated in Fig. 1.4, the beams had a total length of 3200 mm with a 2600 mm span. The 300 mm long overhangs of the beams had a rectangular cross-section, with a 290 mm height

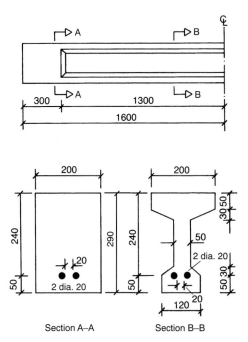

Fig. 1.4. Details of the T-beams designed to the method described in Chapter 4

and a 200 mm width, while the remainder of the beams had a T-section with the shape and dimensions indicated in Fig. 1.4. The longitudinal reinforcement comprised two ribbed 20 mm diameter bars with a yield stress of 500 MPa and an ultimate strength of 670 MPa. The shape and arrangement of the transverse reinforcement is shown in Fig. 1.5. From this figure, it can be seen that the transverse reinforcement comprised two smooth 6 mm diameter bars for beam B and two 1.6 mm diameter wires for beam C, with yield stresses of 570 MPa and 360 MPa respectively. The uniaxial cylinder compressive strength of concrete was 32 MPa. Finally, the loading arrangement used for the tests is shown schematically in Fig. 1.6.

By using the above design characteristics, current design methods[1.6,1.7] predict for both beams a flexural capacity of 72·8 kNm, while the shear capacity is 74 kN for beam B under two-point loading and 20·2 kN for beam C under four-point loading. The value of the total external load corresponding to flexural capacity is 182 kN for the case of two-point loading and 208 kN for the case of four-point loading. On the other hand, the values of the total load corresponding to shear capacity are 148 kN and 40·4 kN for beams B and C, respectively. The smaller of the values of the loads corresponding to flexural and shear capacities for each beam represents the beam load-carrying capacity, i.e. the load-carrying capacities of beams B and C are 148 kN and 40·4 kN respectively. In both cases, the values of load-carrying capacity correspond to shear capacity.

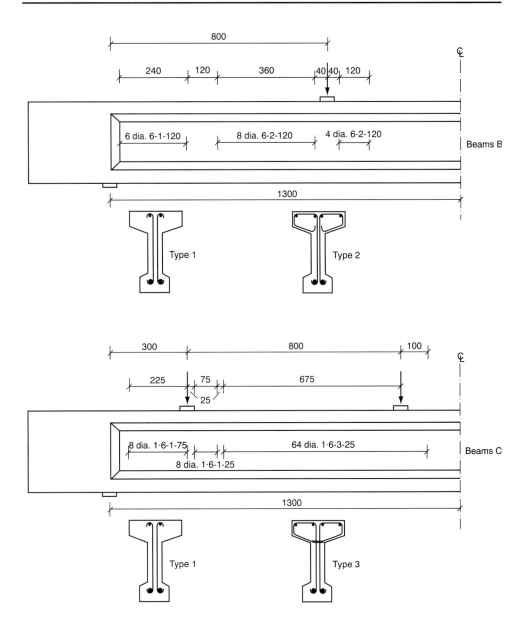

Fig. 1.5. Transverse reinforcement details for the beam in Fig. 1.4 for the case of two-point (beams B) and four-point (beams C) loading

Yet, the load-carrying capacity established by experiment was found to be significantly higher than that predicted for both load cases. In fact, the measured values were found to exceed those corresponding to flexural capacity; for the case of beam B under two-point loading, the load-carrying capacity was measured at 192 kN, while for the case of beam C under four-point loading, it was found to be 240 kN, with both beams exhibiting significant ductility. The predictions for the values of load-carrying capacity and for the modes of failure, together with their experimental counterparts, are summarised in Table 1.1.

From the table, it becomes apparent that, as for the case of the structural walls presented in section 1.2, current design methods are incapable of predicting not only the load-carrying capacity of the beams considered in the present section, but also their mode of failure. Again, the causes for this inability appear to be associated with the methods used to assess shear capacity, for which the size of the deviation of the predicted values from their measured counterparts appears to depend on the loading arrangement. For the case of the two-point loading exerted on beam B, the deviation is of the order of 30%, while for the case of

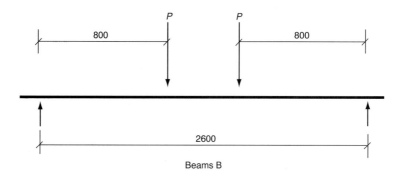

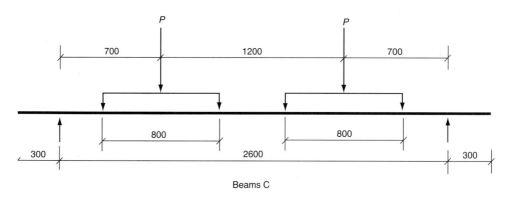

Fig. 1.6. *Schematic representation of loading arrangement used for the beams in Figs 1.4 and 1.5*

Table 1.1. *Comparison between predicted and experimental ultimate loads and failure modes for beams B and C*

Beam	Prediction		Experiment	
	Load-carrying capacity: kN	Mode of failure	Load-carrying capacity: kN	Mode of failure
B	148	brittle	192	ductile
C	40·4	brittle	240	ductile

the four-point loading exerted on beam C, the deviation reaches 500%. Deviations of the size of the latter value cannot be considered as representing an uncommon case, since the bending-moment and shear-force diagrams resulting from four-point loading are similar to those corresponding to uniformly distributed loading, which is the commonest type of loading in practical design.

On the other hand, if one ignores the predictions for 'shear' failure, the predicted values of the load corresponding to flexural capacity appear to deviate from the measured values by approximately 15%. Although such values of the deviation do not justify a radical reappraisal of the concepts underlying the assessment of flexural capacity, an investigation of the causes of such deviations is necessary in order to improve the predictions of flexural capacity.

In contrast with flexural capacity, the huge deviation of the measured value of the load-carrying capacity corresponding to 'shear' failure from its predicted counterpart is indicative of the urgent need for a radical reappraisal of the methods currently used for assessing shear capacity. Such a reappraisal should be based on the investigation of the causes dictating the observed failure characteristics, with the aim of identifying the strength reserves of structural concrete, since, as for the case of the structural wall considered in the preceding section, the experimental results reported in the present section for the beams indicate that the contribution of concrete to shear capacity is significantly larger than that presently allowed for in design. A realistic assessment of this contribution should be expected to lead to a reduction in the amount of transverse reinforcement and, hence, a more efficient design solution.

1.4. Structural wall under combined vertical and horizontal loading

The deviations of the predicted from the experimental values of load-carrying capacity for the structural members discussed in the preceding sections give the impression that the methods currently used in practical design underestimate load-carrying capacity, and, hence, although potentially inefficient or uneconomical, lead to conservative design solutions. In both of the studies considered, the causes of the deviations were found to be associated with the method employed to assess shear capacity, which appears to underestimate the contribution of concrete, thus leading to predictions of load-carrying capacity considerably smaller than the values established by experiment.

The experimental information presented in this section was extracted from reference 1.9 and refers to an RC structural wall subjected to the combined action of vertical and horizontal loading. As shown in Fig. 1.7, the structural wall had a rectangular shape with a height of 1200 mm, a width of 1180 mm, and a thickness of 100 mm. As for the structural walls in section 1.2, the

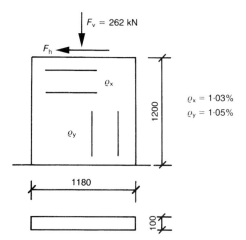

Fig. 1.7. Design details of the structural wall under the combined action of axial and shear edge loading

wall was monolithically connected with an upper and a lower beam, the lower beam being used to clamp the wall to the laboratory floor so as to form a fully-fixed connection, while the combined action of the vertical and horizontal loads was exerted as indicated schematically in Fig. 1.7.

The wall reinforcement comprised 8 mm diameter ribbed bars with values of the yield stress and ultimate strength equal to 574 MPa and 764 MPa respectively. It was uniformly distributed in both the horizontal and vertical dimensions, the percentages of vertical and horizontal reinforcement being 1·05% and 1·03% respectively (see Fig. 1.7). The uniaxial cylinder compressive strength of the concrete used was 35 MPa.

By using the above design characteristics, the value of flexural capacity in the presence of an axial compressive force $V = 262$ kN, as predicted by the method adopted by the European code of practice (Greek version[1.7]), was found to be approximately 570 kNm, the shear capacity (in compliance with the same code) being 753 kN. The external horizontal load, which, when exerted concurrently with a vertical load of 262 kN (see Fig. 1.7), causes a bending moment at the wall base equal to the flexural capacity, may easily be calculated to be equal to 430 kN. If flexural failure could be prevented at the wall base, then the above external horizontal load would increase to a value of 753 kN, which would then cause 'shear' failure. Hence, the methods adopted by current codes of practice predict flexural failure when the horizontal load, combined with a vertical load of 262 kN, attains a value of 430 kN (in Fig. 1.8, the latter value is indicated by the dashed horizontal line (assuming elastoplastic steel behaviour, but the value increases to 460 kN if strain hardening is allowed for) while the analytical curve refers to formal finite-element (FE) modelling[1.10]).

Figure 1.9 depicts the crack pattern of the wall just after failure. From this crack pattern, it becomes apparent that, with the exception of *one* flexural crack which formed at the wall

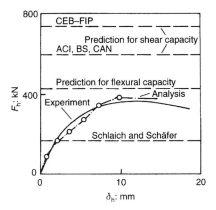

Fig. 1.8. Experimental load–deflection curve and code predictions for the load-carrying capacity of the structural wall in Fig. 1.7

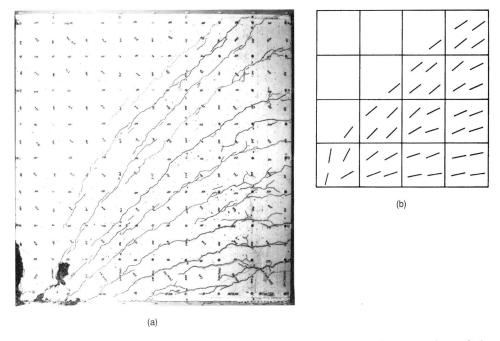

Fig. 1.9. Crack pattern of the structural wall in Fig. 1.7 at failure: (a) experimentally established, and (b) predicted by FE analysis

base, the cracks are inclined. Moreover, the extension of the single flexural crack was interrupted by one of the diagonal cracks which penetrated deeply into the compressive zone and led to total disruption of concrete at the lower left-hand corner of the wall. Therefore, in contrast with the code prediction for flexural capacity, the failure mode of the wall has all the characteristics of a 'shear' type of failure.

It should be noted that the value of the horizontal load causing failure does not exceed 392 kN (see Fig. 1.8), as compared with the value of 753 kN which is the predicted value of shear capacity. The deviation of the prediction from the experimental value is of the order 100%, and, in contrast with the structural members discussed in the preceding sections, this indicates an unsafe design solution since the predicted value is significantly larger than its experimental counterpart. Moreover, it was surprising to note that, while in the preceding cases the current design methods appeared to underestimate the strength reserves of structural concrete in shear, in the present case, not only are such reserves non-existent but the contribution of the reinforcement is also considerably lower than that expected.

The causes of the above deviation cannot be explained on the basis of the concepts which underlie current design methods. Their investigation will form the subject of the next chapter.

1.5. Column with additional reinforcement against seismic action

The column considered in this section is one of the structural members used to investigate the validity of the earthquake-resistant design clauses of the Greek version of the European code of practice for structural-concrete design.[1.11] The earthquake-resistant design clauses recommend additional transverse reinforcement within specific portions of beam-like elements referred to as 'critical lengths'. A characteristic difference between the present column and the structural elements which formed the subject of the preceding sections lies in the loading arrangement which, in the present case, was developed so as to introduce a point of contraflexure within the column length. Such a loading arrangement was considered to be more representative of the loading conditions developing in real frame-like structures where the formation of points of contraflexure is inevitable.

1.5.1. Design details

The experimental set-up employed for the introduction of a point of contraflexure within the column length is depicted in Fig. 1.10.[1.11] The column has a free height of 1000 mm and an orthogonal cross-section 230 mm × 100 mm. The column ends are encased in two reinforced-concrete prisms with a cross-section 400 mm × 100 mm (the latter dimension coinciding with, and aligned in, the smaller column dimension) and a height of 200 mm. The lower prism was used for clamping the column to the laboratory floor in a manner that simulated fixed-end boundary conditions at the column base. In contrast, the upper prism was clamped to a very stiff inverse U-shaped steel frame through which the horizontal load was exerted in the direction of the larger dimension of the column cross-section. The exerting of the horizontal load through the steel frame resulted in a combination of horizontal force and bending moment at the column upper end which, in turn, led to the formation of a point of contraflexure at a distance of 600 mm from the column base.

The reinforcement details are depicted in Fig. 1.10. The vertical reinforcement comprised four 14 mm diameter ribbed steel bars arranged symmetrically about the longitudinal axis at the four corners of the column; the stress–strain curves describing

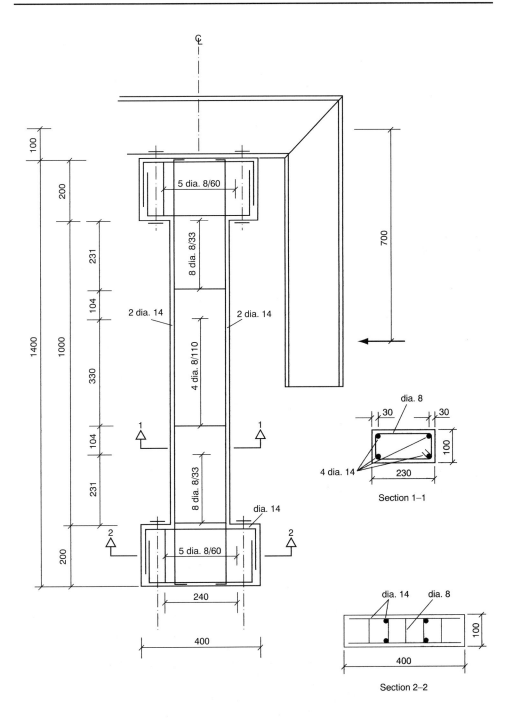

Fig. 1.10. Column design details and loading arrangement

the characteristics of the steel bars are shown in Fig. 1.11. The transverse reinforcement comprised 8mm diameter smooth stirrups; the mechanical characteristics described by the stress-strain curve are also shown in Fig. 1.11. In accordance with the earthquake-resistant design clauses of the Greek version of the

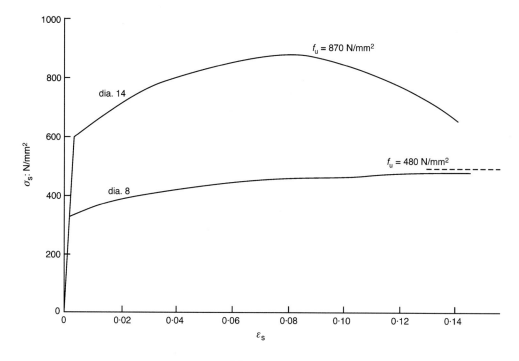

Fig. 1.11. Stress–strain curves of the reinforcement used for column in Fig. 1.10 and beam in Fig. 1.15

European code,[1.7] within the 'critical lengths' which extend to a distance equal to 230 mm from the column ends, the stirrups were arranged with a spacing significantly denser than the value of 110 mm required to safeguard against 'shear' failure. The stirrup-spacing within these lengths was 33 mm, while in the remainder of the column it was 110 mm. The uniaxial cylinder compressive strength of the concrete used for the columns was 30 MPa.

1.5.2. Experimental behaviour

By using the design details shown in Figs 1.10 and 1.11 and assuming all safety factors to be equal to 1, current codes predict a flexural capacity of the column cross-section of 33·41 kNm. Moreover, equating the external load to the internal actions at the column base yields a value of 55·68 kN for the external load causing flexural failure. (This load is equivalent to the combined action of a horizontal load of 55·68 kN and a bending moment of 22·27 kNm at the upper end of the column (see Fig. 1.12(a).)

The bending-moment and shear-force diagrams at the ultimate limit state in flexure are shown (by the full lines) in Fig. 1.12(b), with the former showing the point of inflexion at 600 mm from the column base. The shear-force diagram also includes the values of shear capacity at the 'critical lengths' of the column with the denser stirrup spacing (i.e. $V_u = 190$ kN, which is more than three times larger than the value of the shear force at flexural failure). From the above internal-force diagrams, it becomes apparent that the value of the external load required to cause

PREDICTION VALIDITY APPRAISAL OF CURRENT DESIGN METHODS

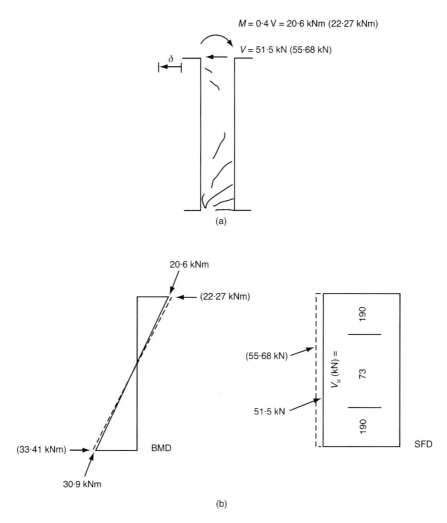

*Fig. 1.12.
Experimental results for the column in Fig. 1.10:[1.11] (a) schematic representation of crack pattern at failure; (b) bending-moment and shear-force diagrams corresponding to load-carrying capacity established experimentally (continuous lines) and flexural capacity (dashed lines) (the shear-force diagram also includes the code values of shear capacity at the various portions of the column)*

'shear' failure is significantly larger than that causing flexural failure. Hence, in accordance with current design methods, the column load-carrying capacity is that corresponding to flexural capacity and it has a value of applied horizontal force equal to 55·68 kN.

Figure 1.12(b) also depicts the bending-moment and shear-force diagrams corresponding to the value of load-carrying capacity established by experiment. The latter value of horizontal force is 51·5 kN against the predicted value of 55·68 kN, i.e. the design prediction overestimates the true column load-carrying capacity by approximately 7%.

Much more worrying, however, it should be noted that, while the prediction refers to flexural failure, the column failed in 'shear,' in contrast with the code prediction for a shear capacity corresponding to a load significantly larger than that causing flexural failure. It should also be noted that 'shear' failure occurred within the lower 'critical length' of the column which was designed to the earthquake-resistant design clauses of the code[1.7] so as to have a shear capacity larger than that of the remainder of the column. In fact, as indicated in Fig. 1.12(b), the deviation of the value of 190 kN, predicted as shear capacity of the 'critical length,' from the value of 51·5 kN, established by experiment, is of the order of 260%, i.e. it is approximately three times larger than the deviation (both in magnitude and in the lack of safety) established for the case of the structural walls discussed in the preceding section.

The above deviation of the predicted from the experimentally established values is indicative of the inadequacy of the concepts which underlie the earthquake-resistant structural-concrete design clauses of current codes of practice. As discussed previously, the above clauses specify additional transverse reinforcement in 'critical lengths' of structural-concrete members, in excess of that required to safeguard against 'shear' failure; the purpose of such additional reinforcement is not only to increase the margin of safety against 'shear' failure, but also to ensure ductility. Yet, the experimental information presented in this section demonstrates clearly that the increase in transverse reinforcement does not achieve either of the above two aims (i.e. improved strength and ductility) of structural design; in fact, it appears to *cause* a brittle, rather than the predicted ductile, mode of failure.

1.5.3. Causes of failure

In contrast with the case of the structural wall of Fig. 1.7, for which there is not sufficient published information to allow the investigation of the causes of premature failure, an attempt to identify the causes of brittle failure for the case of the column presently studied may be based on the concepts which are currently used to describe structural-concrete behaviour. In accordance with such concepts, a beam-like RC structural member with both longitudinal and transverse reinforcement (such as, for example, the column considered in the present section), after the formation of inclined cracking, functions as a truss with the transverse reinforcement (stirrups) and cracked concrete in the tensile zone forming transverse ties and inclined struts, while the compressive zone and the longitudinal reinforcement form the compression and tension chords respectively.

For the column shown in Fig. 1.10,[1.11] just before failure, the width of inclined cracking, which formed at the base of this structural member, was found to be approximately 2 mm. As such a crack width is significantly larger than that considered to allow through 'aggregate interlock' — the transfer of forces across the crack surfaces,[1.12] the inclined struts of the truss can form only within concrete between consecutive inclined or flexural cracks.

The strength of the struts restricts the load-carrying capacity of the truss to a value which is given by the expression $V_{Rd2} = 0.5 f_c b d / \mu$, where f_c is the uniaxial cylinder compressive strength, b and d are the web-width and depth, respectively, of the cross-section, and μ is the safety factor usually taken equal to 2. As the strict quality control enforced in the testing of the columns justifies the use of $\mu = 1$, substituting f_c, b, and d with their values into the above formula yielded $V_{Rd2} = 190$ kN.

An assessment of the forces sustained by the stirrups requires the following: (a) the number of stirrups intersecting an inclined crack at the column base; (b) the width of the crack. In the present case, the 2 mm wide inclined crack at the column base intersected three stirrups.[1.11] As the length of the stirrup was 200 mm, the 2 mm crack width induces in the reinforcement a tensile strain of $\varepsilon = 2/200 = 0.01$ which, as indicated by the $\sigma - \varepsilon$ curve of Fig. 1.11, corresponds to a stress $\sigma_s = 370$ MPa. Hence, the tensile force sustained by the stirrups is $V_s = 111.5$ kN.

Figure 1.13 shows the internal transverse forces developing at a cross-section at the column base including an inclined crack, together with the shear force corresponding to this cross-section. Equating internal to external actions yielded the shear force induced in the concrete (just before column failure), since the tensile force sustained by the stirrups was already assessed as discussed previously, and the shear force corresponding to failure was experimentally determined. From the figure, it can be seen that the size of the tensile force sustained by the stirrups is such (i.e. so large) that its action is resisted by the shear force sustained by concrete instead of the converse being true, i.e. the stirrup taking up some of the shear force (as originally intended). It is also interesting to note that, as the crack width is significantly larger than that allowing the development of 'aggregate interlock', the shear force sustained by concrete is

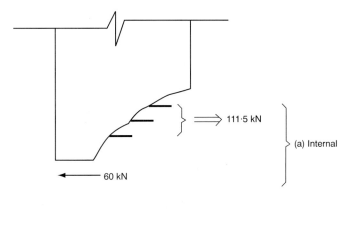

Fig. 1.13. External and internal actions acting on a cross-section, of the column in Fig. 1.10, including a crack with large width

sustained essentially by the compressive zone alone. In fact, the size of the shear force acting within the compressive zone is such that, in accordance with current design concepts, it can only cause 'shear' failure of this zone which, in turn, leads inevitably to collapse of the column.

It would appear from the above that the tensile straining of the stirrups is not always due to the portion of the shear force which, in accordance with the concepts underlying current codes, is resisted by the stirrups. It would seem that there are cases, such as the case of the present column, for which the shear force is not necessarily related to the tensile stress developing within the stirrups. As a result, a unique relationship between shear force and stirrup straining cannot always yield realistic predictions of shear resistance.

1.6. Simply-supported beams with one overhang

It is usually accepted that, when contrasting theoretical predictions, *valid* experimental information — even when stemming from just one test — is sufficient to invalidate a theory, with the term 'valid' referring to the requirement for reliable measurements under definable boundary conditions throughout the test. However, such a requirement is unlikely ever to be fulfilled owing to the secondary testing-procedure effects that are unavoidable in the testing of structural concrete elements, particularly as regards the boundary conditions imposed, which invariably deviate from those intended, with the valid information being that corresponding to the smallest possible deviation from the idealized conditions. As it could be argued that, for the columns discussed in the preceding section, the conflict between the behaviour established by experiment and that predicted by current design concepts may primarily reflect secondary testing-procedure effects, it was considered essential to investigate the validity of the column results by using a completely different experimental set-up for testing another structural-concrete element. The structural element used for the investigation was a simply-supported beam with an overhang tested by means of the experimental set-up shown in Fig. 1.14.

1.6.1. Design details

Figure 1.15 shows that the beam had cross-sectional characteristics similar to those of the column discussed in the preceding section in respect of the dimensions, the type and arrangement of the longitudinal reinforcement, and the type of transverse reinforcement. The transverse reinforcement within the portion of the beam between the supports was arranged in compliance with current code provisions for shear design, while within the overhang the stirrup spacing was that specified by the earthquake-resistant design clauses of the code.[1.7] The material characteristics for both concrete and steel were those for the column in the preceding section.

PREDICTION VALIDITY APPRAISAL OF CURRENT DESIGN METHODS

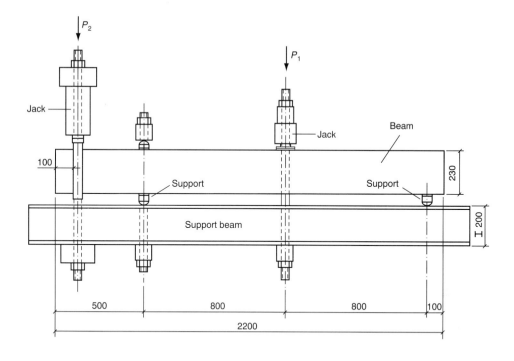

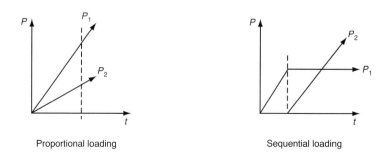

Fig. 1.14. Beam loading arrangement and histories

As indicated by the experimental set-up in Fig. 1.14, the beam was subjected to the combined action of two point loads: P_1 (at the middle of the beam portion between the supports), and P_2 (at a distance of 100 mm from the overhang end). The two loads were combined so as to yield two types of loading regime: *(a)* a proportional loading with $P_1 = 3P_2$; *(b)* a sequential loading in which P_1 was increased to a value of 70 kN, remaining constant at this value thereafter, while P_2 increased monotonically up to the failure of the beam. A schematic representation of the above loading regimes is included in Fig. 1.14.

1.6.2. Experimental behaviour

The main results obtained from the two test types are depicted in Figs 1.16 and 1.17 which show: *(a)* a schematic representation of the beam mode of failure; *(b)* the bending-moment and shear-

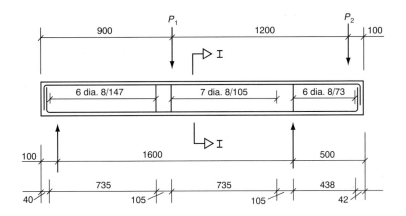

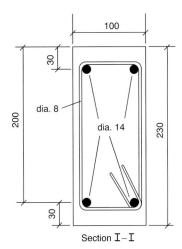

Fig. 1.15. Design details of the beam in Fig. 1.14 under two-point loading

force diagrams which correspond to both the predicted and experimentally established values of the load-carrying capacity; (c) a typical load–deflection curve established by experiment (as at least two tests were carried out for each load case). The predicted values of bending moment and shear force are indicated in parentheses, while the shear-force diagrams also include the predicted values of shear capacity at all the portions of the beam defined by the relevant stirrup spacing.

From the bending-moment and shear-force diagrams, it is apparent that the predicted value of the load-carrying capacity corresponding to flexural capacity is considerably smaller than the predicted load-carrying capacity corresponding to shear capacity. Hence, the predicted load-carrying capacity of the beam is (for both the cases of proportional and sequential loading) that corresponding to flexural capacity. Yet, for both

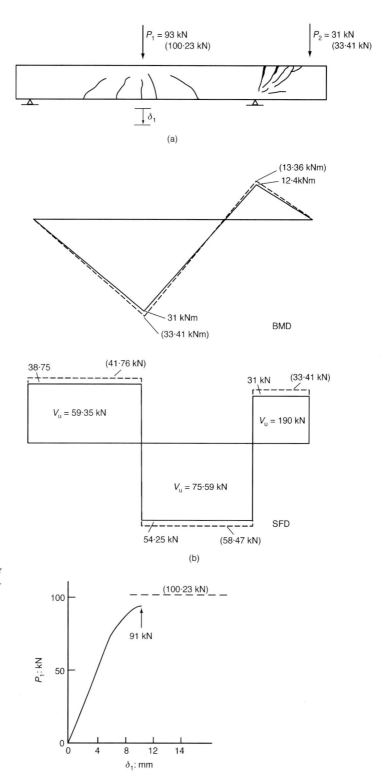

Fig. 1.16. Experimental results for the beam in Fig. 1.15 under proportional loading:[1.11] (a) schematic representation of mode of failure; (b) bending-moment and shear-force diagrams corresponding to the experimentally established load-carrying capacity (continuous line) and flexural capacity (dashed line) (the shear-force diagram includes the shear capacities of the various beam portions); (c) load–deflection curve

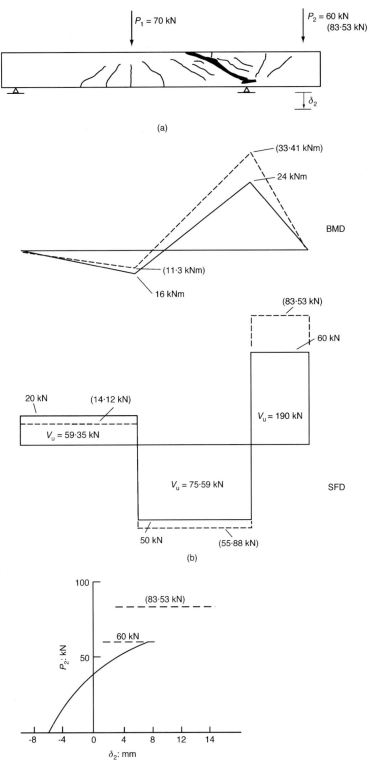

Fig. 1.17. Experimental results for the beam in Fig. 1.15 under sequential loading:[1.11] *(a) schematic representation of mode of failure; (b) bending-moment and shear-force diagrams corresponding to the experimentally established load-carrying capacity (continuous line) and flexural capacity (dashed line) (the shear-force diagram includes the shear capacities of the various beam portions); (c) load–deflection curve*

loading regimes, the beam failed before the flexural capacity was exhausted.

For the case of proportional loading (see Fig. 1.16), failure appears to be associated with the occurrence of severe inclined cracking of the overhang in the region of the support. It is interesting to note that, at failure, both the bending moment and the shear force in the region of the support close to the overhang were equal to approximately 30% of their predicted maximum values. In fact, in accordance with current design concepts, the overhang was the least likely portion of the beam to exhaust its shear capacity, since it not only had the highest shear capacity, but also it was subjected to the smaller shear force.

For the case of sequential loading (see Fig. 1.17), failure occurs within the portion of the beam between the supports within the region closest to the overhang. This region suffered multiple inclined cracking which extended both towards the upper face, in the region of the point load, and towards the support. In contrast with the proportional-loading regime, this region of the beam was the most critical to exhaust its shear capacity, although the applied shear force was significantly smaller than the shear capacity.

1.6.3. Causes of failure

As for the case of the column in Fig. 1.10, the causes of premature (in accordance with current design concepts) 'shear' failure appear to be associated with the width of the inclined cracks which, just before failure occurs, was measured as a little larger than 2 mm. From Fig. 1.16, it becomes apparent that (for the beam subjected to proportional loading) at least three stirrups (with a 200 mm leg length) are intersected by one inclined crack. Hence, the stirrup legs undergo a relative extension $\varepsilon = 2/200 = 0.01$ which, as indicated by the $\sigma - \varepsilon$ curve for dia. 8 bars in Fig. 1.11, corresponds to a stress $\sigma_s = 370$ MPa. As a result, the shear force sustained by the stirrups is $V_s = 3(2\pi 8^2/4)370 = 111.5$ kN $< V_{Rd2} = 190$ kN (for the calculation of the latter, see section 1.5.3).

It would appear, therefore, that beam failure is due to failure of the compressive zone under the action of the large shear force required to maintain the balance between internal and external transverse actions at the cross-section including the 2 mm wide inclined crack (see Fig. 1.18). As for the case of the column (see Fig. 1.13), the compressive zone resists the action of the tensile force sustained by the stirrups rather than the action of the applied shear force. The magnitude of this tensile force is such that it leads inevitably to failure of the compressive zone, thus leading to collapse of the beam.

For the case of sequential loading, although the wide inclined cracks formed in a different region of the beam (see Fig. 1.17),

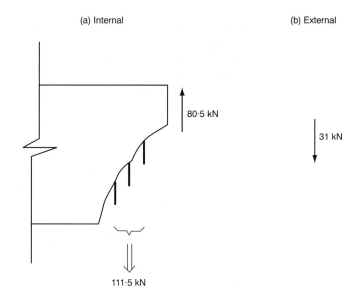

Fig. 1.18. External and internal actions acting at a cross-section, of the beam in Fig. 1.15, including a crack with a large width

the mode of failure of the member appears to remain the same, i.e. the large width of inclined cracking is the underlying cause of failure. It should be noted, however, that the present description of the causes of failure such as the above is based on the concepts which underlie current design methods: therefore, the unexpectedness of the resulting conclusions suggests that a reappraisal of these concepts may lead to a more realistic description of the causes of failure, and, also, contribute to either the improvement of current methods or the development of new methods for practical structural design.

1.7. Failure of structural concrete under seismic load

The Kobe earthquake has been one of the most destructive earthquakes that has ever stricken a populated area. Among the large number of structures devastated by this earthquake, many have been RC structures. The main cause of structural collapse has been the unexpectedly large seismic acceleration which was significantly larger than that used for the earthquake-resistant design of the structures.[1.13] Such an unexpectedly large acceleration caused the development of actions which exhausted all margins of safety allowed for by the design process used; hence, structural collapse was inevitable.

Many of the structures which collapsed during the earthquake were designed in compliance with the 'permissible stress' concept. This concept considers the state of a structure under service loading conditions, and completely ignores the ultimate limit-state characteristics of the structure. As a result, the design method employed was incapable of securing ductile — and, hence, was incapable of safeguarding against brittle — types of failure. In fact, the earthquake action caused a wide range of modes of failure (including flexural, shear, and member-connection failures), which occurred in a manner ranging from brittle to ductile.

From the available data in the literature on the design details and behaviour of particular structures in Kobe during the 1995 earthquake, only a tentative evaluation of the ability of current design concepts to predict structural behaviour may be attempted. Such an attempt may be made by reference to the Higashi–Nada motorway fly-over which collapsed by overturning sideways as shown in Fig. 1.19.[1.14] The figure indicates that collapse was caused by failure of the piers which, in all but one case, appear to have suffered a ductile flexural mode of failure.

An assessment of the flexural and shear capacities of the piers may be carried out by using the design details indicated in Fig. 1.20.[1.15] The piers, with a circular (3100 mm) diameter cross-section and an average height of 11 m, were reinforced in the longitudinal direction with 35 mm diameter bars arranged along the perimeters of two concentric circles, close to the outer perimeter of the pier, each perimeter being formed by 60 equally-spaced bars. (Not all the bars are shown in the cross-sectional view in Fig. 1.20 which, thus, gives only a schematic description of the main reinforcement.) The transverse reinforcement consisted of (two) 16 mm diameter bars forming circular stirrups, with 200 mm spacing, encompassing the (two) rows of longitudinal reinforcement. The lower (2500 mm high) portion of the piers had an additional (third) arrangement of longitudinal reinforcement, similar to the two described above, with circular stirrups also encompassing this additional row of longitudinal bars. The yield stress of all types of reinforcement was $f_y = 400$ MPa, while the uniaxial cylinder compressive strength of concrete was $f_c = 35$ MPa.

Using the above design details, current design methods yield values of the flexural capacities of the sections with the two and three rows of longitudinal reinforcement equal to 65 MNm and 95 MNm respectively, the most critical section being that at a distance of 250 mm from the pier's fixed end. Ignoring, for the moment, the possibility of types of failure other than flexural, the piers should fail in flexure when the transverse load acting at the top end attains a value of 7·65 MN.

For the case of monotonic loading — where both concrete and transverse reinforcement are considered to contribute to shear capacity — current design methods yield a value of shear capacity equal to 12 MN, which is significantly larger than the value of the shear force corresponding to flexural capacity. As a result, under monotonic loading conditions, the piers should fail in flexure before the shear capacity is attained. On the other hand, for the case of load reversals such as those exerted on the piers

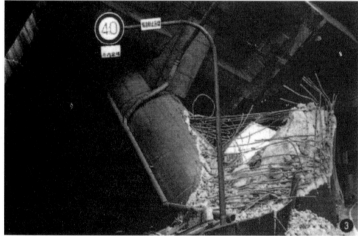

Fig. 1.19. Characteristic failures of the piers of the Higashi-Nada bridge caused by the Kobe earthquake

PREDICTION VALIDITY APPRAISAL OF CURRENT DESIGN METHODS

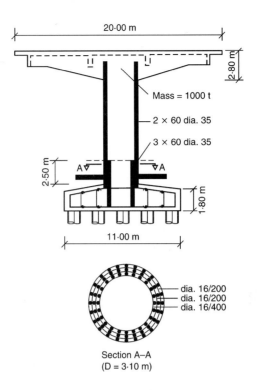

Fig. 1.20. Design details of a typical pier of the Higashi-Nada bridge

under the earthquake action, current shear design methods (in the absence of axial compressive load) ignore the contribution of concrete; hence, the predicted value of shear capacity is 4·8 MN, i.e. it is merely 55% of the shear force corresponding to flexural capacity. Therefore, the piers under cyclic loading conditions (including load reversals) should exhibit a 'shear' type of failure before their flexural capacity is attained.

Figure 1.19 indicates that, in all but only one case, the piers under earthquake action failed in a flexural manner. It would appear from the above, therefore, that, by adopting the reasoning underlying current design methods, the conclusion should be drawn that the piers collapsed under the monotonically increasing load of the first cycle of the earthquake action.

However, the first few cycles of the earthquake action were shown to be insufficient to cause failure of the piers.[1.15] In fact, the ground acceleration exceeded the design value after a number of loading cycles sufficiently large to justify the assumption of current design methods for a negligible contribution of concrete to shear capacity under load reversals. As discussed earlier, for a negligible contribution of concrete, the calculated value of the shear capacity is considerably reduced to a level significantly smaller than the value of the shear force corresponding to flexural capacity. As a result, the piers under earthquake action should have failed in 'shear,' rather than flexure. Yet, as Fig. 1.19

shows, it was the flexural, rather than the shear capacity that was first exhausted. Such behaviour is indicative of the inability of current design methods to yield a realistic assessment of the shear capacity; in fact, the code prediction underestimated the actual value of shear capacity by a margin at least as large as the predicted value. It would appear, therefore, that the additional transverse reinforcement specified in current codes to compensate for the 'assumed' inability of concrete to contribute to shear capacity would, in the present case, be unnecessary.

1.8. Conclusions In all cases presented in this chapter (with the exception of the first case study, namely the structural wall SWA), the deviations of the predicted from the true (experimental) structural behaviour were considerable with regard to both the mode of failure and the load-carrying capacity. In most cases, failure was not the predicted ductile behaviour but a brittle one; while in certain cases, the predicted values of load-carrying capacity were found either to overestimate or to underestimate their experimental counterparts by a large margin.

The above deviations were found to be due primarily to the inability of current design concepts to yield realistic predictions of shear capacity and, to a lesser extent, to the methods used to assess flexural capacity. The magnitude of the deviations was found to be so large that it could only be attributed to the inadequacy of the theoretical basis of current design methods. This inadequacy also becomes apparent from the application of the earthquake-resistant design provisions of current codes which, in spite of the larger number and denser spacing of the stirrups specified in the 'critical lengths' of the structural members investigated, were found to cause brittle failure, and not, as intended — and, indeed, predicted — to safeguard against it while, at the same time, supposedly improving the member's ductility.

There is an increasing body of evidence that points to this need for improving the accuracy of certain code provisions, especially those pertaining to the problem of 'shear' in structural concrete members. For example, the potential lack of safety related to punching failures (misleadingly still referred to usually as 'punching shear') around column-slab zones was stressed in reference 1.10 in 1995 and, whatever the actual causes which led to the subsequent sudden collapse, on 21 March 1997, of part of the Pipers Row car park in Wolverhampton (and the consequent closure of many similar structures), the available photograph and accompanying caption ('Midlands slab punches out warning')[1.16] are certainly not incompatible with this type of failure. Similarly, the case of the $180 million Sleipner offshore platform (totally destroyed on 23 August 1991),[1.17] illustrates how codes may fail to provide accurate predictions of collapse load and/or failure mode (this particular case study will be discussed in Chapter 4).

In view of the above, it is, therefore, not surprising that, in his recent award-winning article, [1.18] Professor Priestley devotes a section to what he rightly describes as 'the shear myth(s)'. Significantly, he states: 'Shear design of reinforced concrete is so full of myths, fallacies, and contradictions that it is hard to know where to begin in an examination of current design. Perhaps the basic myth central to our inconsistencies in shear design is that of shear itself.' (p.61). The remainder of the present book will attempt to remove some of these myths, fallacies and contradictions from design practice.

1.9. References

1.1. Lefas I.D. *Behaviour of reinforced concrete walls and its implication for ultimate limit-state design.* University of London (Imperial College), 1988, PhD thesis.

1.2. Lefas I.D., Kotsovos M.D., and Ambrasseys N.N. Behaviour of reinforced concrete structural walls: Strength, deformation characteristics, and failure mechanism. *ACI Structural Journal*, 1990, **87**, No. 1, January–February, 23–31.

1.3. British Standards Institution. *Structural use of concrete Part 1. Code of practice for design and construction.* BSI, London, 1985, BS 8110.

1.4. American Concrete Institute. *Building code requirements for reinforced concrete.* ACI, Detroit, 1986, ACI 318–83 (revised 1986).

1.5. *Code for the design of concrete structures for buildings.* CSA, Rexdale, 1984, CSA A23.3-M84.

1.6. Comité Euro-International du Béton, CEB-FIP Model code for concrete structures. CEB, Lausanne, 1978.

1.7. Technical Chamber of Greece. *Code of practice for the design and construction of reinforced concrete structures.* TCG, 1991 (in Greek).

1.8. Kotsovos M.D. and Lefas I.D. Behaviour of reinforced concrete beams designed in compliance with the concept of the compressive force path. *ACI Structural Journal*, 1990, **87**, No. 2, March–April, 127–139.

1.9. Maier J. and Thürlimann B. *Bruchversuche und Stahlbetonscheiben.* Institut für Baustatik und Konstruktion, Eidgenössische Technische Hochschule, Zurich, 1985.

1.10. Kotsovos M.D. and Pavlović M.N. *Structural concrete. Finite element analysis for limit-state design.* Thomas Telford, London, 1995.

1.11. Kotsovos M.D., Bazes S., and Lefas I.D. Contribution to the investigation of the validity of the new code for the design of structural concrete. *11th Greek Concrete Congress*, Corfu, May 1994 (in Greek).

1.12. Reinhardt H.W. and Walraven J.C. Cracks in concrete subject to shear. *Journal of the Structural Division, Proc. ASCE*, 1982, **108**, No. ST1, January, 207–224.

1.13. Gazetas G., Michaelides O., Loukakis K., and Mylonakis G. *The 17/1/1995 earthquake in Kobe (Japan): review, comparisons, and preliminary conclusions.* Information Bulletin No. 1851, Technical Chamber of Greece, 27 March 1995, 72–75 (in Greek).

1.14. Japan Society of Engineers. Preliminary Report on *The great Hanshin earthquake.* Japan Society of Civil Engineers, 1995.

1.15. Michaelidis O. and Gazetas G. Dynamic analysis of the failure of the piers of the Higashi-Nada fly-over during the Kobe earthquake. *11th Greek Concrete Congress*, Larnaka, October 1996, **III**, 237–247 (in Greek).

1.16. Parker D. Shock collapse sparks lift slab fears. *New Civil Engineer*, 1997, 27 March/3 April, 3.

1.17. Collins M.P., Vecchio F.J., Selby R.G. and Gupta P.R. The failure of an offshore platform. *Concrete International*, 1997, August, 28–35.

1.18. Priestley M.J.N. Myths and fallacies in earthquake engineering: conflicts between design and reality. *Concrete International*, 1997, February, 54–63.

2. Reappraisal of current methods for structural concrete design

2.1. Introduction

It appears from the data presented in Chapter 1 that the excessive deviations of the predictions of current design methods from the experimentally established behaviour of structural-concrete beam-like members are due to the inadequacy of the theoretical basis of the methods currently used for assessing the flexural and, in particular, the shear capacities of such members. In this chapter, the above methods are concisely described and the results of research carried out to date by the authors on the validity of the concepts underlying these methods are reviewed, in an attempt to identify the fundamental causes which dictate the observed behaviour of structural concrete.

2.2. Current design methods
2.2.1. Physical model

In accordance with the *simplified beam theory*, the internal state of stress of a beam-like member (such as, for example, a simply-supported beam, with a rectangular cross-section, under two-point loading symmetrically arranged with respect to the mid cross-section) may easily be established by using the bending-moment and shear-force diagrams as depicted in Fig. 2.1. The figure indicates that the state of stress (σ, τ) at any point A of the beam may be assessed from the expressions $\sigma = (M/EI)y$ and $\tau = VS/(bI)$, by using the values of the bending moment M and shear force V at the cross-section including point A (where E is the modulus of elasticity, I is the second moment of area of the cross-section, y is the distance of point A from the neutral axis, S is the moment of the shaded area with respect to the neutral axis, and b is the width of the cross-section at the level of point A). Expressing the state of stress (σ, τ), in the form of principal stresses (σ_1, σ_2), and assessing these principal stresses at a sufficiently large number of points, leads to the construction of the principal-stress trajectories depicted in Fig. 2.2, with the full and dashed curves representing the trajectories of the compressive and tensile stresses respectively.

For the case of a plain-concrete beam, owing to the small tensile strength of the material, the beam will crack in regions where the value of the tensile stresses exceeds the material strength. The cracks will form in the direction orthogonal to that of the tensile stresses and, hence, they practically coincide with the compressive stress trajectories.[1.10, 2.1–2.3] The beam load-carrying capacity may increase beyond the value dictated by the tensile strength of concrete if steel bars are placed along the tensile stress trajectories in a number sufficient to sustain the

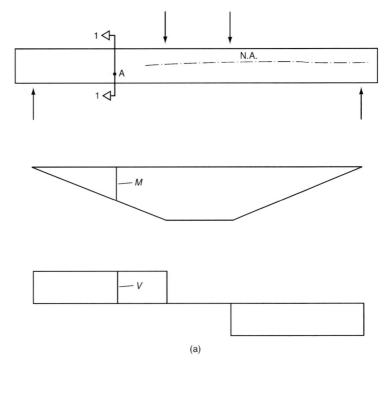

Fig. 2.1. Simply-supported beam under two-point loading: (a) Bending-moment and shear-force diagrams, and (b) distributions of normal (σ) and shear (τ) stresses at cross-section 1-1

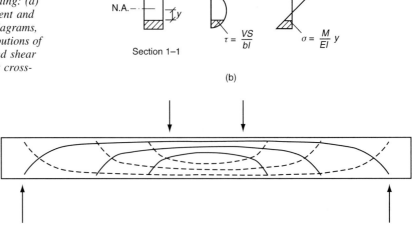

Fig. 2.2. Schematic representation of compressive (continuous lines) and tensile (dashed lines) stress trajectories developing in the beam of Fig. 2.1

portion of the internal tensile actions that cannot be sustained by concrete alone (see Fig. 2.3).

A reinforcement arrangement such as that shown in Fig. 2.3 is, of course, impractical. Instead, in practical design, use is made of straight steel bars placed in both the longitudinal and the transverse directions, as indicated in Fig. 2.4. The figure indicates that the longitudinal bars are placed as close as possible to the

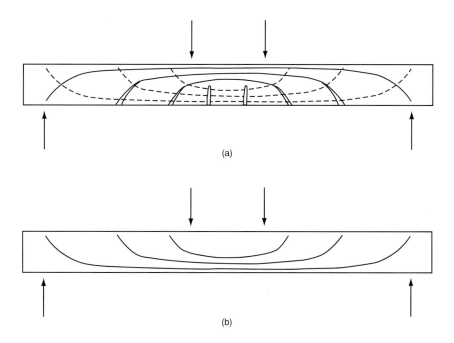

Fig. 2.3. (a) Schematic representation of cracking leading to failure of a concrete beam, and (b) 'theoretical' reinforcement arrangement capable of preventing failure

tensile face of the beam, and they are designed so as to sustain the entire tensile force that develops as a result of the bending action. In contrast with the longitudinal bars, the transverse bars are distributed within the 'shear spans' of the beam, and they are designed so as to sustain the portion (V_s) of the applied shear force (V_a) in excess of that that can be sustained by concrete alone (V_c). With such a reinforcement arrangement the 'shear spans' are considered to function as *trusses* after the formation of inclined cracks: the compressive zone and the longitudinal bars form the compression and tension *chords* of the trusses respectively, while the transverse bars form the *ties,* and the cracked concrete within the tensile zone — through the 'aggregate interlock' that develops at the inclined crack surfaces — allows the formation of inclined *struts*. (The *truss* concept is discussed in more detail in section 2.2.3.)

The physical model of Fig. 2.4, which is a combination of *beam* (within the 'flexure span') and *truss* (within the 'shear

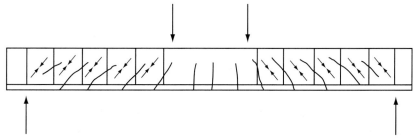

Fig. 2.4. Practical reinforcement arrangement of a concrete beam

spans'), underlies the methods currently used for structural-concrete design. The model was employed initially within the context of the *permissible-stress philosophy* which assumes linear-elastic behaviour, while the design methods were later modified so as to enable their use for the case of non-linear behaviour which characterises both the materials and the structure at their *ultimate-limit state*. The model essentially functions as a beam before the occurrence of inclined cracking; however, it is progressively transformed into a *truss* in the regions of inclined cracking, while the middle portion of the beam continues to behave as a *beam*.

2.2.2. Beam action

Figure 2.5 depicts a cracked simply-supported reinforced-concrete (RC) beam under the action of transverse uniformly distributed load (q), together with the bending-moment (M) and shear-force (V) diagrams. In the figure, the shaded portion between consecutive flexural cracks in the middle region of the beam will be considered. For the equilibrium of the shaded portion, *beam action* requires the change (ΔM) in bending moment to balance the action of the couple of shear forces (V), and, concurrently, the change (ΔV) in shear force to balance the external load $q\Delta x$ (see Fig. 2.5(a)). Figure 2.5(c) also shows a schematic representation of the internal actions (which develop in concrete and steel), so as to satisfy the equilibrium conditions of the shaded portion. It should also be noted that, in the zone of this portion, the shear force is significantly lower than that required to cause 'shear' failure and, hence, the beam requires only a nominal amount of stirrups in this zone of the beam.

The extension of the *simplified beam theory* for the case of non-linear behaviour — which characterises not only concrete and steel, but also beam behaviour at the ultimate limit state — is based on the following assumptions.[2.4]

(a) Plane cross-sections remain plane during bending (Bernoulli assumption).
(b) Concrete behaviour in the compressive zone is adequately described by an uniaxial axial stress–axial strain curve. This curve together with assumption *(a)* define fully the stress distribution in the compressive zone (see Fig. 2.5(c)).
(c) The tensile strength of concrete is ignored.
(d) The behaviour of the steel bars is described by the stress–strain curves for steel in compression or tension.
(e) Failure occurs when the strain of concrete at the extreme compressive fibre attains a critical value ε_u usually taken to be equal to 0·0035.
(f) There is full bond between concrete and steel.

Assumptions *(a)* and *(f)* satisfy the compatibility conditions, while assumptions *(b)* to *(e)* define material behaviour. The

REAPPRAISAL OF METHODS FOR STRUCTURAL CONCRETE DESIGN

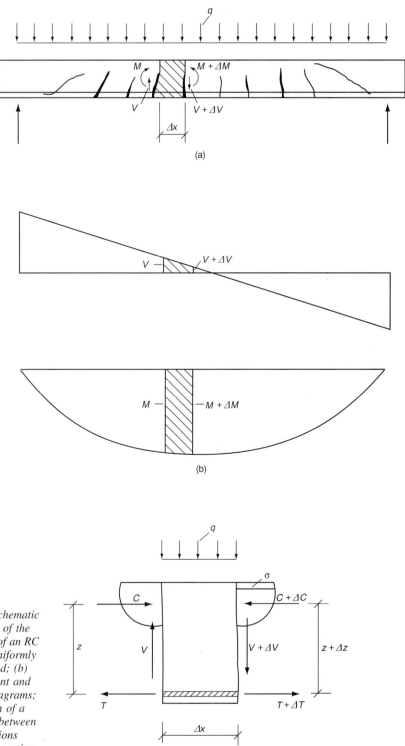

Fig. 2.5. (a) Schematic representation of the crack pattern of an RC beam under uniformly distributed load; (b) bending-moment and shear-force diagrams; (c) equilibrium of a beam portion between two cross-sections including consecutive cracks

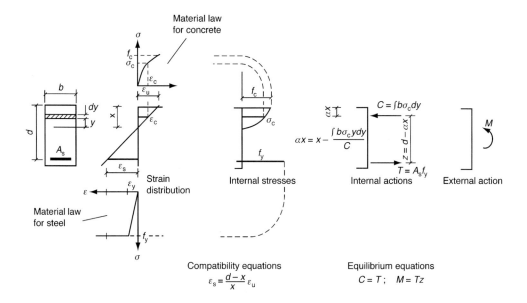

Fig. 2.6. Concepts and conditions of equilibrium and compatibility underlying the method of calculation of flexural capacity

application of the above assumptions to the equilibrium conditions yields analytical relations which express the internal actions as functions of the cross-sectional geometric (b, d, A_s) and material characteristics (f_c, f_y). A concise presentation of these relations, together with a schematic representation of the geometric characteristics and the external and internal actions they correlate, is provided in Fig. 2.6. When designing a beam in flexure (within the context of the *limit-state philosophy*), this involves the assessment of the geometric characteristics so that the expressions of Fig. 2.6 are satisfied.

2.2.3. Truss action

As mentioned in section 2.2.1, after the formation of inclined cracks, the presence of transverse reinforcement is considered to transform the beam into a *truss*, with this reinforcement essentially forming the transverse ties of the truss.[2.4] The opening of the inclined cracks causes an elongation of the transverse reinforcement, and, hence, induces in it an internal tensile force. Moreover, since transverse reinforcement is anchored in regions close to the upper and lower faces of the beam, the above tensile force is transferred to the entire beam web, causing (in accordance with the statical requirements of the *truss analogy*) the formation of diagonal struts through cracked concrete, the latter constituting the largest portion of the beam web, as depicted in Fig. 2.4. However, for cracked concrete to allow the formation of diagonal struts, it must be feasible for a compressive force to be transferred across the surfaces of inclined cracks. Such a transfer is considered to be effected through the *interlocking of the aggregates* at the surfaces of inclined cracks; *aggregate interlock* resists the shear displacement of these surfaces, which is also resisted by *dowel action*, i.e. the shear stiffness of the longitudinal reinforcement which crosses the inclined cracks close to the tensile face of the beam. The truss is considered to be completed by the compressive zone and the longitudinal tensile reinforcement which form, respectively, the compressive and tensile chords of the truss depicted in Fig. 2.4.

The application of the *truss analogy* in practical structural-concrete design is effected through the following assumptions.[2.4]

(a) 'Shear' failure occurs when the shear force acting at a *critical cross-section* attains a limiting value related to the shear capacity of concrete. (For a beam with the same cross-sectional characteristics throughout its span, the *critical cross-section* is the cross-section under the largest shear force, ignoring the end portions of the beam extending to a distance equal to the beam depth from the supports, where the shear force is considered to be directly transferred to the supports by *arch action*.)

(b) Of the shear force acting at a critical cross-section (which includes an inclined crack), a portion of it is sustained by concrete, while the remainder is sustained by the longitudinal and transverse reinforcement intersecting the inclined crack. The shear capacity of the critical cross-section is attained when both concrete and transverse reinforcement attain their strength values in shear and tension respectively, while a portion of shear capacity is contributed by the longitudinal reinforcement through *dowel action*.

(c) The shear capacity of concrete is due to the shear strength of uncracked concrete in the compressive zone and the contribution of *aggregate interlock* which is effected by the shearing movement of the inclined crack surfaces. In fact, *aggregate interlock* makes up the largest part — ranging from 40% to 70% — of the shear capacity of a critical cross-section without transverse reinforcement, with uncracked concrete and *dowel action* each making (similar) contributions ranging from 15% to 30%.

(d) The compatibility conditions between the elongation of the transverse reinforcement and the crack width are ignored.

The above assumptions are used to evaluate two limiting values forming lower (V_{Rd1}) and upper (V_{Rd3}) limits to the shear capacity of the truss, with the two limits corresponding to the tensile force that can be sustained by the smaller number of transverse links allowed (nominal transverse reinforcement) and the compressive strength of the diagonal struts respectively. For a design shear force V_{Rd2} smaller than V_{Rd1}, the amount of nominal transverse reinforcement specified should be capable of sustaining V_{Rd1}. For a value of V_{Rd2} such that $V_{Rd1} < V_{Rd2} < V_{Rd3}$, a portion of V_{Rd2} equal to V_{cd} is considered to be sustained by the cross-section without the contribution of transverse reinforcement, while transverse reinforcement is specified in an amount sufficient to sustain the remaining portion of the shear force $V_{wd} = V_{Rd2} - V_{cd}$. For $V_{Rd2} > V_{Rd3}$, the cross-sectional dimensions should be modified so that the condition $V_{Rd2} < V_{Rd3}$ is satisfied. Transverse reinforcement usually comprises two-legged stirrups with spacing $s < 0.7d$ (where d is the cross-section depth). Assuming that the longitudinal projection of the inclined crack has a length of $0.9d$, the total stirrup cross-section required within this length for sustaining V_{wd} is easily obtained by $A_{sw} = (s/0.9d)(V_{wd}/f_{yw})$ (where f_{yw} is the yield stress of the stirrups). The assessment of A_{sw} forms the main objective of the shear design procedure. Analytical expressions that may be used to assess V_{Rd1}, V_{Rd3}, and V_{cd} may be found in current codes such as, for example, references 1.6 and 1.7.

2.3. Reappraisal of the current approach for assessing flexural capacity

From among the assumptions presented in section 2.2.2, assumptions *(b)* to *(d)*, which describe the non-linear behaviour of concrete and steel at the material level, provided the basis for extending the use of the *simplified elastic beam theory* to the description of the non-linear beam behaviour at the ultimate-limit state. However, while the material model adopted for the description of steel behaviour is widely considered to be satisfactory, doubt has already been expressed regarding the validity of the model adopted for describing concrete behaviour.[1.10, 2.5–2.8] The doubts expressed concern the validity, on the one hand, of the experimental information used for describing the post-peak stress–strain characteristics of concrete,[1.10, 2.5, 2.7, 2.8] and, on the other hand, the use of *uniaxial* stress–strain characteristics for the description of the deformational response of an element of concrete in the compressive zone of a beam-like structural member in flexure as the ultimate load is approached.[1.10, 2.9] In fact, it has been shown that the transverse expansion of concrete as the peak of its stress–strain curved is neared — which is completely ignored in current design — causes the development of a complex triaxial stress field which essentially dictates the failure mechanism of the structural member.[1.10, 2.9]

It would appear from the above, therefore, that there is a need for a reappraisal of the material model used to describe the behaviour of concrete in the compressive zone of a beam in flexure. Such a reappraisal should include not only the $\sigma - \varepsilon$ relationships, but also the failure mechanism of beam-like members; its aim should be to improve the theoretical basis of structural concrete design so as to be compatible with concrete behaviour as described by valid experimental information.

2.3.1. Concrete behaviour

2.3.1.1. Stress–strain curves

The behaviour of concrete in uniaxial compression is described by $\sigma - \varepsilon$ curves such as those in Fig. 2.7, which depict the relationships between compressive stress (σ) and strains along (ε_a), and transversely to (ε_t), the direction of σ. Figure 2.7 depicts also the relationships between σ and volumetric strain ($\varepsilon_v = \varepsilon_a + 2\varepsilon_t$) for concrete. A characteristic feature of the curves of Fig. 2.7 is that they comprise two branches: an ascending branch and a descending one. It should be noted that current methods for assessing flexural capacity make use of the $\sigma - \varepsilon_a$ curve only (henceforth $\sigma - \varepsilon$ for brevity) for the calculation of the force sustained by the compressive zone.

It is well-known that the $\sigma - \varepsilon$ curves describing the behaviour of concrete in uniaxial compression are obtained from tests on concrete specimens, such as, for example, cylinders or prisms, loaded through steel platens. However, the difference in the mechanical properties between concrete and steel inevitably causes the development of frictional forces at the specimen/platen interfaces. These forces restrain the lateral expansion of concrete at the end zones of the specimen, and, hence, modify the intended stress conditions in these zones.

Although one of the main objectives of current testing techniques is the elimination of the above frictional forces, this objective has proved impossible to achieve to date.[2.8] Figure 2.8 shows characteristic $\sigma - \varepsilon$ curves established from tests on cylinders in uniaxial compression by using various techniques for reducing friction at the specimen/platen interfaces.[1.10, 2.5] From the figure, it can be seen that, in contrast with the ascending branch, which is essentially independent of the technique used to reduce friction, the slope of the descending branch increases with the efficiency of the friction-reducing medium employed. In fact, the increase in slope is such that it leads to the conclusion that, if it were possible to eliminate friction entirely, the descending

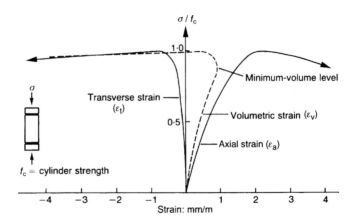

Fig. 2.7. Experimental stress–strain curves for concrete in uniaxial compression

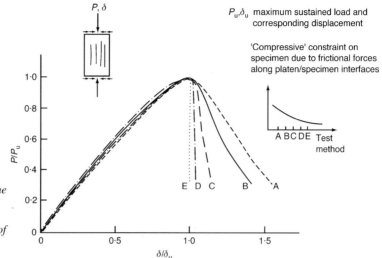

Fig. 2.8. Effect of frictional forces at the specimen/platen interfaces on the stress–strain curves of concrete in compression

branch would have a 90° slope, which is indicative of an immediate and complete loss of load-carrying capacity as soon as the peak stress is attained.

It would appear from the above, therefore, that the descending branch of a $\sigma - \varepsilon$ curve essentially describes the interaction between specimen and the loading platens and *not*, as widely considered, specimen behaviour. Specimen behaviour is described only by the ascending branch of an experimentally established $\sigma - \varepsilon$ curve, and loss of load-carrying capacity occurs in a brittle manner. A similar conclusion may be drawn from experimental information obtained in a recently completed international co-operative project organised by RILEM TC-148 SSC.[2.8]

It may be noted from the $\sigma - \varepsilon_a$ curve of Fig. 2.7 that, at peak stress, ε_a (=0·002) deviates considerably from the value of the axial strain at the extreme compressive fibre of a beam-like member at its ultimate limit state in flexure, which is usually larger than 0·0035.[2.9] Such a deviation is indicative of the inability of uniaxial stress–strain characteristics to describe the true behaviour of concrete in compression within a beam-like member at its ultimate limit state in flexure. Clearly, the stress conditions in the compressive zone are multiaxial, which explains the higher strains attained in actual structural components. The mechanism which leads to the development of such a stress condition is discussed in section 2.3.1.3.

An important characteristic of the behaviour of concrete in compression (which is ignored by current design methods, as implied earlier, when it was stressed that only the $\sigma - \varepsilon_a$ curve, and not the $\sigma - \varepsilon_t$ characteristic, is considered in codes of practice) is the transverse expansion of the material which, when

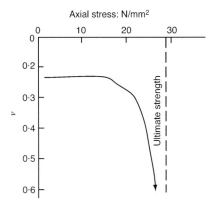

Fig. 2.9. Typical variation of Poisson's ratio with increasing stress obtained from tests on cylinders under uniaxial compression

a level close to the peak stress is attained, begins increasing faster than the material shortening in the direction of the applied compression, in a manner that causes, after an initial continuous consolidation of the material up to this stress level, a considerable volume dilation (see Fig. 2.7). This rapid increase in transverse expansion is also apparent in Fig. 2.9 which describes the variation of Poisson's ratio (ν) with increasing stress (σ). (ν is defined as the ratio of the change in lateral strain ($\Delta\varepsilon_t$) to the change in axial strain ($\Delta\varepsilon_a$).) The figure indicates that the initially constant value of ν increases rapidly beyond a certain value of σ, and, at a stress level close to the peak stress, attains a value of 0.5 which is the limiting value for a continuum. In fact, the value of v at peak stress may even be larger than 1, which is indicative of the discontinuous nature of concrete caused by the cracking processes described in the following section.

2.3.1.2 Cracking

The main cause of the non-linear behaviour of concrete under load is a microcracking process which this material undergoes during loading.[1.10, 2.10] It is generally accepted that the cause of microcracking is the proliferation of flaws which exist within concrete even prior to the application of load. These flaws are attributable to a number of causes, the main ones being:[1.10, 2.2]

   (a) discontinuities in the cement paste owing to its complex morphology
   (b) voids caused by shrinkage or thermal movements as a result of incompatibility between the properties of the various phases present in concrete
   (c) discontinuities at the boundary between aggregate particles and the paste or mortar caused by segregation
   (d) voids present in concrete due to incomplete compaction; etc.

These pre-existing flaws, which may be seen as a constituent of concrete, are randomly distributed and orientated within the material and exhibit a range of shapes and sizes.

The stress or strain state applied to the boundary of an element of a multiphase material such as concrete generates a stress/strain field within the material, which is dependent on the distribution of the component phases (i.e. aggregate particles and cement paste matrix), and the size and distribution of the flaws. Local strain concentrations, therefore, develop throughout the material owing to the incompatible deformation of the constituent phases. Such stress/strain concentrations are further intensified to far higher orders of magnitude because of the presence of flaws, particularly those with high aspect ratios. These flaws are considered to be the potential sources of any load-induced cracking.[1.10, 2.2]

A review[1.10, 2.2] of both theoretical considerations as well as experimental investigations into the behaviour of a microcrack within a stressed brittle material has led to the conclusion that the mechanism of the fracture process — which initiates in the region of one of the above flaws in order to relieve the high tensile stress/strain concentrations developing near the flaw tips — is that of crack extension due to initiation of branches. This process is followed by stable propagation of these branches, as a result of which the process eventually becomes unstable, leading to ultimate collapse. Such crack extension and propagation were found, experimentally, to occur in the direction of the maximum principal stress, with the plane of the propagating crack being orthogonal to the direction of the minimum principal stress (considering compressive stresses as positive).[1.10] Moreover, a propagating crack opens perpendicularly to the crack surface and, hence, causes the formation of voids within the body of the material. Owing to their orientation, such voids affect predominantly the transverse deformation of a concentrically compressed cylinder and lead to the non-linear behaviour indicated in Figs 2.7 and 2.9 discussed in the preceding section.

For the case of a compressive loading, crack extension, in spite of the voids that it causes, does not essentially alter the area of the cross-section orthogonal to the applied load and, hence, it does not affect the load-carrying capacity of this cross-section in a direct manner (see Fig. 2.10(a)). Instead, with the redistribution of the internal stresses that it causes, it not only reduces the tensile stress/strain concentrations existing near the crack tips, but it also increases the energy absorption of the material through the consumption of a portion of the work done by the applied load for fuelling the cracking process. This cracking process continues at the microscopic level of observation until the volume of the material reaches its minimum value (see the $\sigma - \varepsilon_v$ curve in Fig. 2.7). At this point, the capacity of the material for

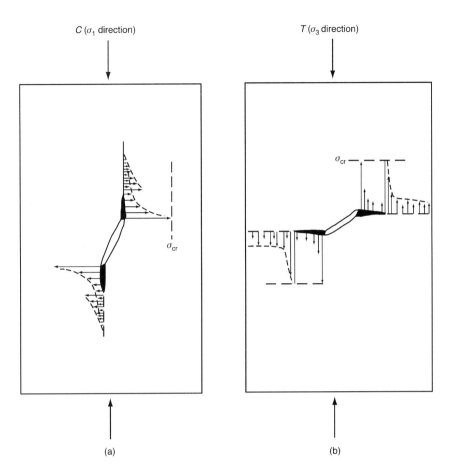

Fig. 2.10. Schematic representation of tensile stress concentrations at the crack tips for the cases of (a) uniaxial compression and (b) uniaxial tension

storing and dissipating energy is exhausted, while a number of microcracks join to form macrocracks (visible cracks), which cause significant volume formation leading to dilation, and almost immediate failure of the material.

For the case of a tensile loading, crack extension reduces the cross-sectional area orthogonal to the applied load and, hence, reduces the load-carrying capacity of this cross-section (see Fig. 2.10(b)). As a result, crack extension cannot cause a redistribution of the internal stresses/strains since it tends to increase, instead of reducing (as for the case of compressive loading), the high tensile stress/strain concentrations and thus leads inevitably to failure of the material. It appears, therefore, that the reduction of the cross-sectional area normal to the applied load is the cause underlying the small tensile strength of concrete.

2.3.1.3. Effect of small transverse stresses on strength and deformation

The assumption that uniaxial stress–strain characteristics are capable of describing the behaviour of concrete in the compressive zone of a beam-like member in flexure implies that the existence of small stresses, acting in directions orthogonal to the longitudinal axis of the member, have an insignificant effect on the material behaviour, and, therefore, can be ignored. Such a simplification, however, contrasts with experimental information such as that shown in Fig. 2.11, which depicts a typical failure envelope for concrete under axisymmetric stress conditions.[1.10, 2.1] (It should be noted that the failure envelope of Fig. 2.11 describes the combination of the principal stresses marking the occurrence of macrocracking which, in contrast with the microcracking process that occurs throughout the loading history and dictates the non-linear constitutive behaviour of concrete, corresponds to the loss of load-carrying capacity.) The figure shows that a small confining stress of the order of $0 \cdot 1 f_c$ (where f_c is the uniaxial cylinder compressive strength of concrete) leads to an increase of the compressive strength in the orthogonal direction by more than 50%. On the other hand, a small tensile stress of the order of $0 \cdot 05 f_c$ appears to be sufficient to reduce the compressive strength in the orthogonal direction by a similar amount.

In fact, the small compressive stresses which develop in the transverse direction of a beam-like member in flexure has been shown to be of the order of $0 \cdot 1 f_c$.[1.10, 2.9] As a result, their effect on the strength of the compressive force is considerable and should not be ignored (while, of course, this important effect is neglected in current design philosophy). Similarly, the effect of the above small stresses on deformation should also not be ignored: they

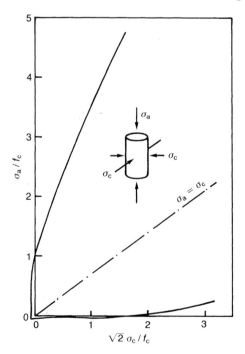

Fig. 2.11. Strength of concrete under axisymmetric stress

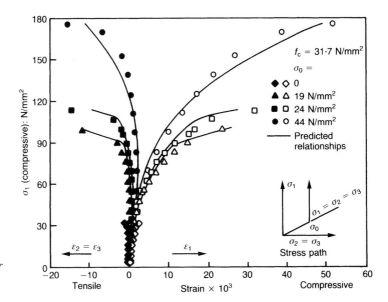

Fig. 2.12. Characteristic stress–strain curves for a typical concrete under axisymmetric compression

create a triaxial stress state in the compressive zone which, as indicated in Fig. 2.12, is capable of inducing in concrete an axial strain significantly larger than 0·0035, which is the limiting value of the strain of the extreme compressive fibre specified by most current codes (and this, only by invoking the dubious premise of the descending branch).

Therefore, it would appear from the above that, by ignoring the considerable effect of the small transverse stresses on strength and deformation of concrete in the compressive zone, current methods were led to combine the use of uniaxial stress–strain characteristics with strain-softening material properties in an attempt to achieve a realistic prediction of the axial strain of concrete at the extreme fibre of the compressive zone. However, the above approach has contributed not only to design guidelines of dubious validity, but also to major confusion — and even disorientation — of research work concerned with the investigation of the behaviour of concrete, at both the material and structure levels. The prominence placed by such research on the importance of the strain-softening characteristics obscures the true underlying causes of the observed structural-concrete behaviour.

2.3.2. Failure mechanism of the compressive zone

2.3.2.1. A fundamental explanation of failure initiation based on triaxial material behaviour

The description of the salient characteristics of concrete at the material level has been presented in the preceding sections in sufficient detail to enable these properties to be implemented in a method suitable for the design of structural-concrete beam-like members. Before this is done, however, it is possible to anticipate the typical mode of failure initiation in such structures by means of a simple reasoning based on two key features of concrete materials. These two fundamental characteristics of concrete are shown schematically in Figs 2.7 and 2.11, which summarise much of the earlier discussion on material behaviour.

As is well known, concrete is weak in tension and strong in compression. Its primary purpose in an RC structural member is to sustain compressive forces, while steel reinforcement is used to cater for tensile actions <u>and concrete provides protection to it</u>. Thus, since the structural role of concrete is concerned primarily with compressive stress states, the present discussion relates to its strength and deformational response under such conditions. Now, as described in section 2.3.1.1, information on the strength and deformational properties of concrete is usually obtained by the testing of cylinder or prism specimens under uniaxial compression. Although a set of typical stress–strain curves stemming from such tests has been discussed in the earlier part of this chapter, it is useful to refer again to Fig. 2.7 which depicts, in a generic sense, the above set of curves. The figure serves as a reminder that, in addition to the strain in the direction of the loading (which usually constitutes the main — if not the sole — item of interest in current design thinking), the uniaxial test also provides information on the strain perpendicular to this direction. Furthermore, a typical plot of volumetric-strain variation appears in the figure. A characteristic feature of the curves in Fig. 2.7 is that they comprise ascending and gradually descending branches. However, despite the prominence given to the latter in design, it was explained in section 2.3.1.1 how experimental evidence shows quite conclusively that, unlike the ascending branch, the descending branch does not represent actual material behaviour: rather, it merely describes secondary testing-procedure effects due to the interaction between testing machine and specimen. This is an important observation concerning the behaviour of concrete at the material level as the lack of strain-softening, i.e. post-ultimate branch, provides justification for it to be referred to as a brittle material. On the other hand, it turns out that considerations of the behaviour of concrete at the structural level make the actual post-ultimate response of the material irrelevant, because, even if the latter were to exist, failure of concrete in a structure occurs invariably prior to the attainment of its ultimate compressive stress. The case for such a statement may be argued along the following lines.

Perhaps the most significant feature of concrete behaviour is the abrupt increase of the rate of lateral expansion a uniaxial test specimen undergoes when the load exceeds a level close to, but not beyond, the peak stress. Such a feature was already noted in section 2.3.1.1, and the relevant stress level may be identified as the minimum-volume level (see Fig. 2.7) which marks the beginning of a dramatic volume dilation that, in the absence of any frictional restraint at the interface between the ends of the specimen and the steel platens, is considered to lead rapidly to failure even if the load remains constant. This is why the stress at which concrete begins to expand is associated with a process governed essentially by void formation which, for all practical purposes, may be equated to the failure load, as explained in section 2.3.1.2. It is important to emphasise here that the rapid expansion at the minimum-volume level, where the tensile strain at right angles to the direction of maximum compressive stress soon exceeds the magnitude of the compressive strain, is a feature of both uniaxial and the more general triaxial compressive behaviour.[1.10, 2.11]

The other key feature of concrete behaviour relates to the major role played by even relatively small (secondary) stresses when assessing the true bearing strength of the material. This was illustrated and fully discussed in section 2.3.1.3 by reference to Fig. 2.11, which indicates schematically the variation of the peak axial compressive stress sustained by cylinders under various levels of confining pressure. Such behaviour implies that the presence of small secondary stresses developing within a structural member in the region of the path along which compressive forces are transmitted to the supports should have a significant effect on the load-carrying capacity of the member: compressive stresses should increase it considerably, whereas tensile stresses should — dramatically — have the opposite effect.

Even though the above two fundamental characteristics of concrete at a material level are well known, they are rarely (if ever) mentioned; more important still from a design viewpoint, their implications for the behaviour of concrete in a structure do not appear yet to have been fully appreciated in terms of failure mechanisms resulting from the interaction of concrete elements in RC structures. In order to appreciate that such interactive behaviour is unavoidable irrespective of the type of structure and/or loading conditions, it is useful to recall that, owing to the heterogeneous nature of concrete, the stress conditions within a concrete structure or member can never be uniform even under uniform boundary conditions. As a result, even for the case of a cylinder subjected to uniform uniaxial compression, the development of triaxial stress conditions is inevitable owing to the setting up of secondary stresses that are essential for maintaining compatibility of deformation within the structure (see Fig. 2.13). Under service loading conditions, the secondary stresses are negligible and can be ignored for design purposes. However, as the load increases, volume dilation occurs in a localised region where the stress conditions are the first to reach the minimum-volume level. Concrete dilation is restrained by the

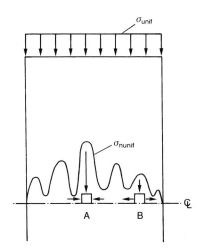

Fig. 2.13. Schematic representation of the non-uniform stress distribution σ_{nunif} due to material heterogeneity within a concrete cylinder under uniform compressive stress σ_{unif} (A: element under triaxial compression; B: element under compression-tension)[1.10]

surrounding concrete, and this is equivalent to the application of a confining pressure which, as Fig. 2.11 indicates, should increase the strength of the dilating region. At the same time, the dilating region induces tensile stresses in the adjacent (restraining) concrete (as every action has its equal and opposite reaction) and, on the basis of the information shown in Fig. 2.11, these should reduce the strength of the concrete.

The preceding reasoning for the failure of a specimen under nominally uniform stress conditions is even more evident in the general case of arbitrary structural systems in which there is always a localised region in compression where the minimum-volume level is exceeded before it is exceeded in surrounding regions (which are also in compression). As a result, the rate of tensile strain will increase abruptly in this region, thus inducing tensile stresses in the adjacent concrete. Concurrently, compatibility and equilibrium require that the surrounding concrete should restrain the expansion of the localised region. While this extra restraint further increases the strength of the localised region, the tensile stresses eventually turn the state of stress in the surrounding concrete into a state of stress with at least one of the principal stress components tensile, and thus they reduce the strength in the latter zone. Therefore, it is always the concrete surrounding the localised region of wholly compressive stresses that fails first, since its state of stress now has at least one tensile principal-stress component.

It appears, therefore, that, owing to the interaction of the concrete elements within a structure, failure is unlikely to occur in regions where the compressive stress is largest. Instead, failure should occur in adjacent regions, where the compressive stresses may be significantly smaller, owing to the presence of small secondary tensile stresses developing as discussed above. Such a failure mechanism indicates that concrete invariably fails in tension, and that a concrete structure collapses before the (usually triaxial) ultimate strength of concrete in compression is exceeded anywhere within the structure.[1.10] This notion that the concrete in the 'critical' zones of compression always fails by 'splitting' — never by 'crushing' — contrasts with widely-held views that form the basis of current analysis and design methods for RC structures. Accordingly, most design procedures have been developed on the assumption that it is sufficient to rely almost entirely on uniaxial (compressive) stress–strain characteristics for the description of concrete behaviour. Invariably this assumption is justified by the fact that structural members are usually designed to carry stresses mainly in one particular direction, and that the stresses that develop in the orthogonal directions are small enough to be assumed negligible for any practical purpose. However, such reasoning underestimates the considerable effect that small stresses have on the load-carrying capacity and on the deformational response of concrete beyond the in-service conditions. The ignoring of these small stresses in design necessarily means that their actual effect on structural behaviour is normally attributed to other causes that are expressed in the form of — as it turns out, erroneous — design assumptions. The following example will suffice to illustrate this. It is often pointed out that the strains recorded in the compressive zone of beams indicate that these are well in excess of that value corresponding to the peak stress in a uniaxial cylinder or prism test. As a result, the argument is put forward that strain softening must be present since such large strains are observed only in the region of the descending branch of the uniaxial test. However, the true explanation lies in the fact that such regions are always subjected to a state of triaxial compression, and this means that, although the peak stress has not been exceeded, the associated triaxial strains are much larger than their uniaxial-test counterparts around the minimum-volume level (as already mentioned in section 2.3.1.3 and illustrated in Fig. 2.12). Thus, for instance, an axial strain of around 8 mm/m marks practically the end of the descending branch for a concrete of $f_c \sim 32$ N/mm^2 in accordance with an ordinary (i.e. uniaxial) cylinder test (Fig. 2.14(a)); when subjected to a hydrostatic stress $\sigma_0 = 24$ N/mm^2, the axial strain for the same concrete prior to the attainment of the descending branch is about three times this value, and becomes much higher still with increasing confinement σ_0 (Fig. 2.14(b)).

2.3.2.2. Triaxiality and failure initiation by macrocracking: some experimental and analytical evidence
The behaviour of structural concrete outlined in the preceding section has been predicted by analysis[1.10, 2.11, 2.12] and verified by experiment.[1.10, 2.9, 2.13] The inevitable triaxiality conditions in zones usually (misguidedly) deemed to be critical on account of

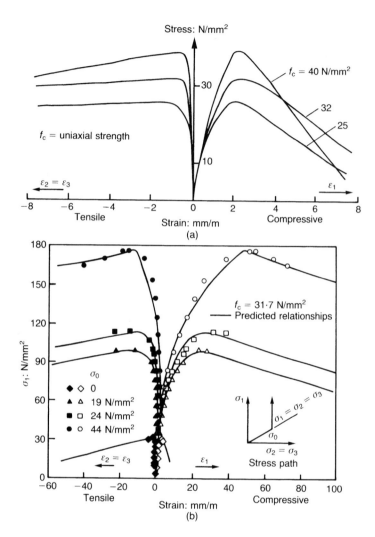

Fig. 2.14. Typical stress–strain relationships for concretes obtained from tests on cylinders:[1.10] (a) various concretes under uniaxial compression; (b) given concrete (with $f_c = 31.7$ MPa) under triaxial compression for various values of hydrostatic stress σ_0

large compressive action, and the associated failure initiation by tensile stresses adjacent to such zones has become evident throughout the various problems tackled by means of finite-element (FE) modelling in reference 1.10. Nevertheless, it is instructive to devote the present section to a preliminary illustration of the basic mechanism that governs the ultimate-load conditions in a concrete member.

An RC beam designed in accordance with typical current regulations based on the ultimate-strength philosophy will be considered. The stress–strain characteristics of concrete in compression are considered to be adequately described by the deformational response of concrete specimens such as prisms or cylinders under uniaxial compression; thus, the ensuing stress distribution in the compressive zone of a cross-section at the ultimate limit state, as proposed, for example, by BS 8110,[1.3, 2.14] exhibits a shape similar to that shown in Fig. 2.15(a). The figure indicates that the longitudinal stress increases with the distance from the neutral axis up to a maximum value and then remains constant. Such a shape of stress distribution has been arrived at on the basis of both safety considerations (with built-in safety factors) and the widely-held view that the stress–strain relationship of concrete in compression consists of both an ascending and a gradually descending portion, as illustrated in Fig. 2.15(b). (In fact, the stress block in Fig. 2.15(a) is based on the simplification that, beyond the peak stress, perfect plasticity may be assumed up to a strain of 0·0035; however, alternative stress blocks may also be used, either involving further simplification such as full plasticity leading to a rectangular stress block[1.3] or derived by allowing for strain softening between peak stress and a strain of 0·0035 so that the shape of the stress block is curved throughout.[2.14]) The portion beyond the ultimate (i.e. peak) stress in Fig. 2.15(b) defines the post-ultimate stress capacity of the material which, as indicated in Fig. 2.15(a), is generally considered to make a major contribution to the maximum load-carrying capacity of the beam. It will be noticed that the principal reasoning behind the stress block adopted for design purposes is based on the large compressive strains (in

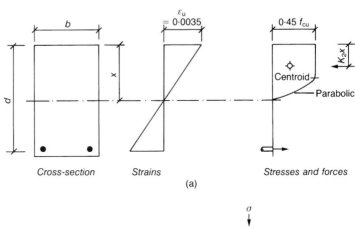

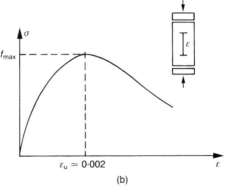

Fig. 2.15. Characteristic design of a beam cross-section for ultimate-load conditions:[2.9] (a) stress and strain distribution proposed by BS 8110 (Part 1) for a critical section at failure (f_{cu} = characteristic cube strength); (b) typical stress–strain relationships for concrete under uniaxial compression used to derive (a) ($f_{max} = f_c$ for cylinders $\approx 0 \cdot 8 f_{cu}$)

excess of 0·0035) measured on the top surface of an RC beam at its ultimate limit state, such strains being almost twice the value of the compressive strain ε_u at the peak-stress level under uniaxial compression. (Typically, ε_u is of the order of 0·002 — see Fig. 2.15(b).)

The above design procedure is not, however, borne out by experimental evidence, which can be shown by reference to the results obtained from a test series of three simply-supported rectangular RC beams subjected to flexure under two-point loading.[2.9] The details of a typical beam are shown in Fig. 2.16, with the central portion under pure flexure constituting one-third of the span. The tension reinforcement consisted of two 6 mm diameter bars with a yield load of 11·8 kN. The bars were bent back at the ends of the beams so as to provide compression reinforcement along the whole length of the shear spans. Compression and tension reinforcement along each shear span were linked by seven 3·2 mm diameter stirrups. Neither compression reinforcement nor stirrups were provided in the central portion of the beams. Owing to the above transverse reinforcement arrangement, all beams failed in flexure rather than shear, although the shear span-to-effective depth ratio was 3. The beams, together with control specimens, were cured under damp hessian at 20° C for seven days and then stored in the laboratory atmosphere (20° C and 40% relative humidity) for about two months, until tested. The cube and cylinder strengths at the time of testing were $f_{cu} = 43·4$ N/mm² and $f_c = 37·8$ N/mm² respectively. Besides the load measurement, the deformational response was recorded by using both 20 mm long electrical resistance strain gauges and linear-voltage displacement transducers (LVDTs). The strain gauges were placed on the top and side surfaces of the beams in the longitudinal and the transverse directions as shown in Fig. 2.17. The figure also indicates the position of the LVDTs which were used to measure deflection at mid-span and at the loaded cross-sections. Finally, the stress–strain characteristics in uniaxial compression for the concrete used in the investigation are depicted in Fig. 2.18.

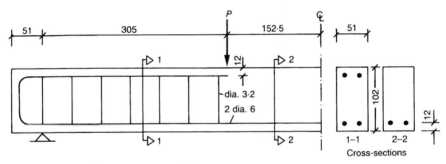

Fig. 2.16. RC beams under two-point loading:[2.9] beam details

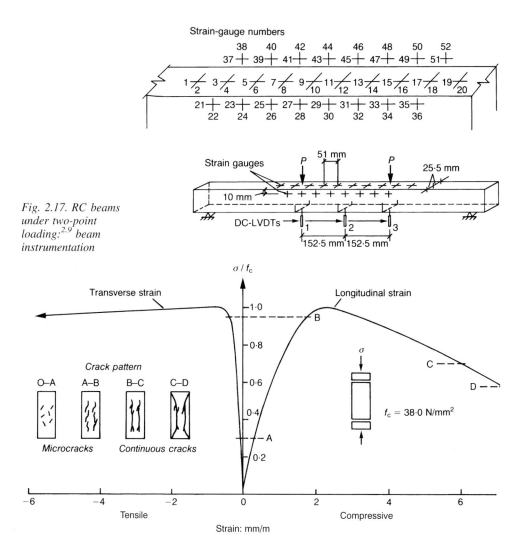

Fig. 2.17. RC beams under two-point loading:[2.9] beam instrumentation

Fig. 2.18. RC beams under two-point loading:[2.9] stress–strain relationships under uniaxial compression for the concrete mix used

In presenting the salient results of the test series of beams, it is convenient to begin by showing the relationships between longitudinal (i.e. along the beam axis) and transverse (i.e. across the beam width) strains, as measured on the top surface of the girders. The relevant information is summarised in Figs 2.19(a) and 2.19(b) which refer to the strains recorded at the critical sections (i.e. throughout the middle third of the beam span) and within the shear spans respectively. Also plotted on these figures is the relationship between longitudinal and transverse strains derived on the basis of the uniaxial material characteristics of Fig. 2.18. Now, if the uniaxial-compression stress–strain characteristics of Fig. 2.18 were to provide a realistic prediction of concrete behaviour in the compressive zone of the beams tested in flexure, then one would expect the relationships

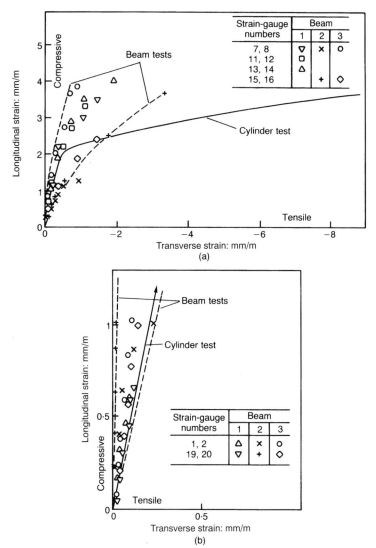

Fig. 2.19.
Relationships between longitudinal and transverse strains measured on the top surface of the RC beams under two-point loading (for strain-gauge locations, see Fig. 2.17):[2.9] (a) at critical sections; (b) within the shear spans

between longitudinal and transverse strains measured on the top surface of the beams to be compatible with their counterparts derived on the basis of the cylinder test; furthermore, longitudinal macrocracks ought to appear on the top surface of the beams, as indicated in Fig. 2.18, where typical crack patterns of axially-compressed concrete cylinders around (B–C) and beyond (C–D) ultimate strength are depicted schematically. It is apparent from Fig. 2.19(a), however, that, for the region of cross-sections including a primary flexural crack, only the portion of the deformational relationship based on the uniaxial cylinder test up to the minimum-volume level can provide a realistic description of the beam's behaviour. Beyond this minimum-volume level, there is a dramatic deviation of the cylinder strains from the beam relationships. Not only does such behaviour support the view that the post-peak branch of the deformational response of a cylinder in compression does not describe material response but, more importantly for present purposes, it clearly proves that, while uniaxial stress–strain data may be useful prior to the attainment of the peak stress, they are insufficient to describe the behaviour once this maximum-stress level is approached. On the other hand, while Fig. 2.19(a) demonstrates the striking incompatibility between cylinder specimen and structural member beyond compressive strains larger than about 0·002 (which, as noted earlier, corresponds to ε_u, the strain at the f_c (or f_{cu}) level — see Figs 2.15 and 2.18), Fig. 2.19(b) shows that the relationships between longitudinal and transverse strains measured on the top surface within the shear span of the beams are adequately described by the longitudinal strain–transverse strain relationship of concrete under uniaxial compression. It should be noted, however, that the relationships of Fig. 2.19(b) correspond to stress levels well below ultimate strength.

An indication of the causes of behaviour described by the relationships of Figs 2.19(a) and 2.19(b) may be seen by reference to Fig. 2.20, which shows the change in shape of the transverse deformation profile of the top surface of beam 1 (but typical of all beams) with load increasing to failure. The characteristic feature of these profiles is that, within the 'critical' central portion of the beam, they all exhibit large local tensile strain concentrations which develop in the compressive regions of the cross-sections where the primary flexural cracks, that eventually cause collapse, occur. Although small strain concentrations may develop in these regions at early load stages before the occurrence of any visible cracking, they become large only when the ultimate limit state is approached and visible flexural cracks appear in the tension zones of the beams. Such a large and sudden increase in transverse expansion near the ultimate load is indicative of volume expansion and shows quite clearly that, even in the absence of stirrups, a triaxial state of stress can be developed in localised regions within the compressive zone. The local transverse expansion is restrained by concrete in adjacent regions (as indicated by the resultant compression forces F in Fig. 2.20), a restraint equivalent to a confining pressure that will later be shown as being equivalent to at least 10% of f_c; hence, as Fig. 2.11 indicates, the compressive region in the plane of a main flexural crack is afforded a considerable increase in strength so that failure is not initiated there. Concurrently, the expanding concrete induces tensile stresses in adjacent regions (these are indicated by the resultant tension forces F and $F/2$ in Fig. 2.20), and this gives rise to a compression/tension state of stress. Such a stress state reduces the strength of concrete in the longitudinal direction, and collapse

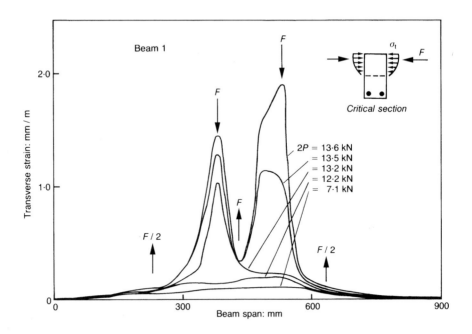

Fig. 2.20. Typical variation of deformation profile of loaded face of RC beams under two-point loading with increasing total load (2P) and schematic representation of resulting forces (F) and stresses $(\sigma_t)^{2.9}$

occurs as a result of horizontal splitting of the compressive zone in regions between primary flexural cracks, as illustrated schematically in Fig. 2.21. Concrete crushing, which is widely considered to be the cause of flexural failure, thus appears to be a *post-failure* phenomenon that occurs in the compression zone of cross-sections containing a primary flexural crack due to loss of restraint previously provided by the adjacent concrete.

It may be concluded from the above, therefore, that the large compressive and tensile strains measured on the top surface of the central portion of the beams should be attributed to a *multiaxial* rather than an uniaxial state of stress. A further indication that these large strains cannot be due to post-ultimate stress–strain characteristics is the lack of any *visible* longitudinal cracking on the top surface for load levels even near the maximum load-carrying capacity of the beams. As shown in Fig. 2.21, such cracks characterise the post-ultimate strength behaviour of concrete under compressive states of stress. Visible cracks occur predominantly on planes parallel to the top surface *at the moment* of final collapse. The typical view of the beam once the collapse of a member has taken place is depicted in Fig. 2.21(b), where the pair of main flexural cracks observed correspond to the peak tensile strain concentrations recorded experimentally in beam 1 (see Fig. 2.20).

It is interesting to note that the results described so far do not contradict the view expressed throughout this chapter that concrete in compression suffers a complete and immediate loss of load-carrying capacity when ultimate strength is exceeded. The implication of the results of the beam tests is that, in the

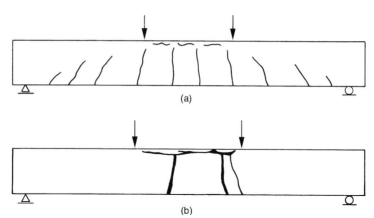

Fig. 2.21. Typical failure mode of RC beams in flexure: (a) a schematic representation of failure mechanism at collapse;[1.10, 2.13] (b) observed failure of tested beams following collapse[2.9]

absence of a post-ultimate gradually falling branch of the stress–strain relationships of concrete in compression, the large compressive strains which characterise RC structures exhibiting 'ductile' behaviour under increasing load (i.e. behaviour characterised by load–deflexion relationships exhibiting trends similar to those shown in Fig. 2.22 for the under-reinforced members tested) are due to a complex multiaxial compressive state of stress which exists in any real structure at its ultimate limit state. Such stress states may be caused by secondary restraints imposed on concrete by steel reinforcement, boundary conditions, surrounding concrete, etc. The significance of these restraints is, in most cases, not understood or simply ignored. It may thus be concluded that the ultimate strength of concrete in localised regions exhibits significant variations dependent on the local *multiaxial* compressive state of stress within the compressive zone of an RC structure or member. The higher the multiaxial ultimate strength of concrete at a critical cross-section, the larger the corresponding compressive and tensile strains. The 'ductility' of the structure, therefore, seems to be dependent on the true (i.e. triaxial) ultimate strength of concrete at critical cross-sections rather than on stress redistributions due to post-ultimate material stress–strain characteristics, even if the latter were assumed to exist.

The state of compressive triaxial stresses compatible with the deformations and strains measured in the beams tested remains to be explored. In addition to the main longitudinal (σ_1) and the secondary transverse (σ_t) stresses, another set of secondary actions also exists, namely the radial stresses (σ_r) acting vertically. Clearly, vertical stresses must exist at, and in the vicinity of, the point loads, but the radial stresses referred to are additional to these and are more relevant for present purposes. These radial stresses are associated with the radial stress resultant (R) which develops within the deformed beam due to the inclination of the compressive (C) and tensile (T) stress resultants

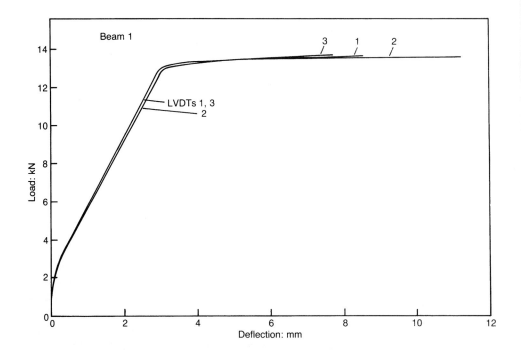

Fig. 2.22. Typical load–deflection curves of RC beams under two-point loading (for LVDT's locations, see Fig. 2.17).[1.10, 2.9]

acting in the longitudinal direction. The above stress resultants are shown schematically in Fig. 2.23, which indicates that even the loaded face (which is generally assumed to be under plane-stress conditions) is subjected to a radial stress resultant. As long as the beam exhibits near-elastic behaviour, the radial stresses corresponding to the radial stress resultant are small in magnitude since they are distributed over the whole length of the central portion of the beam. However, when the central portion of the beam starts to develop large deflections (see Fig. 2.22) due to the formation of a 'plastic' zone caused by a critical flexure crack, the radial stresses become significant in magnitude since they tend to become localised and to concentrate over the plastic zone.[2.15] For load levels close to the maximum load-carrying capacity of the beam, the mean value of the above radial stresses may be estimated — albeit roughly — as follows. If the inclination of the longitudinal compressive and tensile stress resultants is defined by the angle of discontinuity α resulting from the inelastic deformation of the 'plastic' zone (Fig. 2.23), then

$$R = C \sin \alpha = T \sin \alpha \tag{2.1}$$

Now, T is approximately equal to the total yield force of the reinforcement, i.e. $T = 2*11800 = 23600$ N (see earlier details), whereas an approximate value for α may be obtained by the ratio:

$$\alpha/2 = \text{maximum mid-point deflection/half-span of beam} \tag{2.2}$$

Fig. 2.23. Schematic representation of radial actions due to deflected shape of RC beams:[1.10, 2.9]
(a) forces (R);
(b) stresses (σ_r)

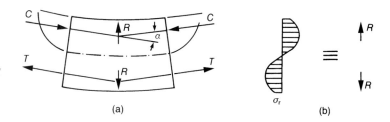

For a value of the maximum mid-point deflection approximately equal to 10 mm (see Fig. 2.22) equation (2.2) gives $\alpha \sim 4\cdot4 \times 10^{-2}$ rad which, when substituted in equation (2.1), results in $R \sim 1000$ N. Finally, assuming that the length of the 'plastic' zone is 5 mm, a nominal value for the radial stresses (approximating the section width to ~ 50 mm) is: $\sigma_r \sim 1000/(5*50) \sim 4$ N/mm^2. Hence, $\sigma_r \sim 0\cdot1*f_c$ since, as noted earlier, $f_c \sim 38$N/mm^2.

The order of magnitude of the transverse stresses σ_t may be assessed by reference to the estimate obtained for σ_r. Consider Fig. 2.24, which shows the variation on the critical section of the average strains measured in the loading direction on the side faces of the beams with the transverse strains measured on the loaded surface. It is interesting to note from the figure that the strains measured on the side faces are slightly larger than those measured on the loaded face. This is considered as an indication that the average value of the stresses restraining the transverse expansion of the critical section should be at least as large as that of the radial stresses, i.e. $\sigma_t > 0\cdot1*f_c$. The transverse and radial stresses, therefore, combined with the longitudinal stresses give rise to a complex multiaxial compressive state of stress in the regions of the large tensile strain concentrations within the compressive zone of the beams. Under such a three-dimensional stress state, concrete can sustain both stresses and strains which can be considerably larger than those obtained in uniaxial material tests that form the basis of most current structural design.

How large are the main stresses σ_1? One would expect these to be at least 50% in excess of f_c since, as pointed out in section 2.3.1.3, Fig. 2.11 suggests that an axisymmetric confining pressure of some 10% of f_c boosts the actual strength by about one-half of its original value. That this is indeed the case may be seen by reference to Fig. 2.25(a), which shows the resultant tension (T) and compression (C) force resultants at a critical section of a beam. Since only an order of magnitude estimate of σ_1 is required, average stress values may be used and hence it is sufficiently accurate to adopt a rectangular stress block. Now, earlier calculations for beam 1 gave $T = C = 23600$ N (i.e. ductile failure), while the ultimate load $P = 6800$ N combined with a rounded-off value of the shear span of some 300 mm leads to the maximum-sustained bending moment of $\sim 6800*300$

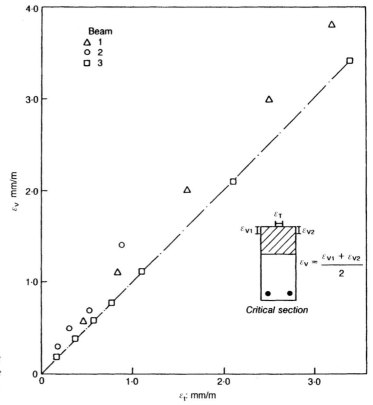

Fig. 2.24. Relationships, with increasing load, between transverse and vertical strains at critical sections of the RC beams under two-point loading[1.10, 2.9]

~ 2040000 Nmm. The lever arm then follows at $z = 2040000/23600 \sim 86 \cdot 5$ mm, enabling the depth of the stress block to be estimated at $x = 2*(90-86 \cdot 5) \sim 7$ mm. As before, the beam width may be approximated to 50 mm so that the compressive-zone stresses $\sigma_1 \sim 23600/(50*7) \sim 67$ N/mm,^{2} i.e. the average value of the longitudinal stress at a critical section is 75% above f_c and, clearly, some of the actual local stresses will be even higher than this figure.

On the basis of the assumed distribution of secondary (i.e. 'confining') stresses σ_t and σ_r (see Figs 2.20 and 2.23), one could argue that the degree of triaxiality varies throughout the depth of the compressive zone in the manner shown in Fig. 2.25(b), with the longitudinal stresses σ_1 increasing from the neutral plane up to a maximum value (where the confinement is greatest) and then gradually decreasing to a smaller value at the loaded face. If so, it might be suggested that — neglecting the inevitable stress variations across the beam width, which only a proper three-dimensional analysis could reveal — the shape of the σ_1 distribution is not unlike that of the generally accepted stress-block shape derived on the basis of a uniaxial stress–strain relationship possessing a gradually descending post-ultimate branch which, as discussed earlier, is used by current

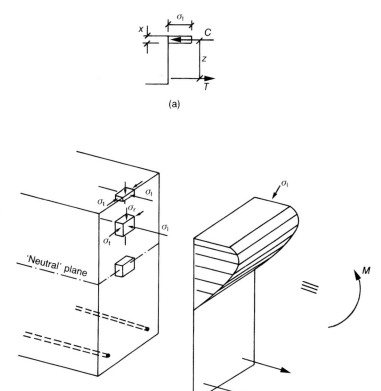

Fig. 2.25. Longitudinal stresses σ_1 in the critical compressive zone of the RC beams under two-point loading at failure: (a) assessment of average σ_1 based on measured values of ultimate tensile force resultant and bending moment;[1.10, 2.13] (b) likely shape of σ_1 distribution predicted on the basis of triaxial behaviour[1.10, 2.9]

design procedures recommended by codes of practice. However, the preceding study shows beyond doubt that, while both the large strains required for ductility and the shape of the stress block might appear as admitting the postulate that uniaxial material properties are applicable at a structural level, such a postulate does not accord with the actual mechanism of failure in a structure and, furthermore, leads to massive underestimates of the true stresses and transverse tensile strains under ultimate conditions. (In view of the latter, it is obvious that the various refinements in the shapes of the stress blocks — see the discussion at the start of the present section — are totally unjustified, so that the simplest stress-block shape (i.e. rectangular) might as well be used in ordinary design calculations.) Therefore, the main conclusion to be drawn from the preceding study is that the importance of triaxiality in elucidating what triggers the collapse of a structure and the sensitivity of triaxial failure envelopes to even small degrees of confinement make it mandatory to incorporate multiaxial material descriptions in any model of analysis aimed at accurate predictions of ultimate behaviour at the structural level.

2.4. Reappraisal of the current approach for assessing shear capacity

The concepts which underlie current shear design methods are incompatible with the behaviour of concrete (at the material level) as described in the preceding section. As an example of this incompatibility, reference could be made to the contribution of cracked concrete — through *aggregate interlock* — to the shear capacity of beam-like structural-concrete members. Such a contribution can be effected only in the presence of the strain-softening characteristics described by the descending branch of a $\sigma - \varepsilon$ curve, since it is this branch that describes the behaviour of concrete after the formation of macrocracks.[2.8] However, from the experimental information discussed in section 2.3.1.1, it became apparent that the descending branch does not describe material behaviour, but merely represents the interaction between specimen and loading platens. Concrete is a brittle material, and, as such, it is characterised by a complete and immediate loss of load-carrying capacity as soon as macrocracking occurs, with such behaviour precluding any direct contribution of cracked concrete to the load-carrying capacity of a structural-concrete member.

Moreover, *aggregate interlock* is considered to be effected by the shearing movement of the interfaces of an inclined crack. However, such a movement is incompatible with the cracking mechanism of concrete discussed in section 2.3.1.2. This cracking mechanism involves crack extension in the direction of the maximum principal compressive stress and crack-opening in the orthogonal direction. The apparent lack of shearing movement of the crack interfaces also contrasts with the assumption that *dowel action* is one the contributors to the shear capacity of a beam-like structural-concrete member.

The doubts expressed above regarding the ability of cracked concrete to contribute to shear capacity also cast doubts on the validity of the concepts of *truss analogy* and *shear capacity of critical cross-section*, as the validity of the above concepts depends on the validity of the concepts of *aggregate interlock* and *dowel action*. Therefore, the objective of the experimental information presented in what follows has been not only to provide definitive conclusions regarding the validity of the concepts which underlie current methods for shear design, but also to identify both the *true* contributors to shear capacity and the causes which underlie the so-called 'shear-types' of failure.

2.4.1. Validity of concepts underlying shear design

The validity of the concepts which underlie current methods for shear design has been investigated experimentally by testing under two-point loading the simply-supported beams shown in Figs 2.26(a) and 2.26(b).[2.16, 2.17] The figures depict the geometric characteristics, together with the reinforcement details, of two types of beam with values of the shear span-to-depth ratio (a_v/d) approximately equal to 1·5 and 3·3, respectively. The beams of

Fig. 2.26(a) have the same geometric characteristics and longitudinal reinforcement but, with regard to the transverse reinforcement, they have been classified as follows:

- beam A: without transverse reinforcement
- beam B: with transverse reinforcement within the shear span only
- beam C: with transverse reinforcement throughout the beam span
- beam D: with transverse reinforcement within the flexure span in the region of the point loads only.

As for the case of the beams in Fig. 2.26(a), the beams of Fig. 2.26(b) also have similar geometric characteristics and longitudinal reinforcement, and, depending on the arrangement of their transverse reinforcement, have been classified as follows:

- beam A1: without transverse reinforcement
- beam B1: with transverse reinforcement within the shear span only
- beam C1: with transverse reinforcement within the portion of the shear span extending to a distance equal to 200 mm from the support
- beam D1: with transverse reinforcement within the portion of the shear span between the cross-section at a distance equal to 200 mm from the support and the cross-section through the point load.

The experimental results are summarised in Figs 2.27(a) and 2.27(b) which show the load-deflection curves for the beams tested.

2.4.1.1. Shear capacity of critical cross-section
In accordance with current design provisions,[1,3] the flexural capacity of the beams in Fig. 2.26(a) is approximately equal to 2·64 kNm and corresponds to values of the load-carrying capacity equal to 39 kN for the case of the beams with $a_v/d = 1.5$ (in Fig. 2.26(a)), and 18 kN for the case of the beams with $a_v/d = 3.3$ (in Fig. 2.26(b)). Moreover, the shear capacity of the cross-section without the contribution of the transverse reinforcement is approximately 7·8 kN, and corresponds to a value of the load-carrying capacity equal to 15·6 kN for both types of beam tested. It should also be noted that the transverse reinforcement is sufficient, in accordance with current code provisions, to safeguard against 'shear' types of failure within the portions of the beams where it was placed.

In accordance with the assumption of the *critical cross-section*, every cross-section within the shear spans of the beams is potentially *critical*, while those within the portions without transverse reinforcement have the smallest shear capacity. Since,

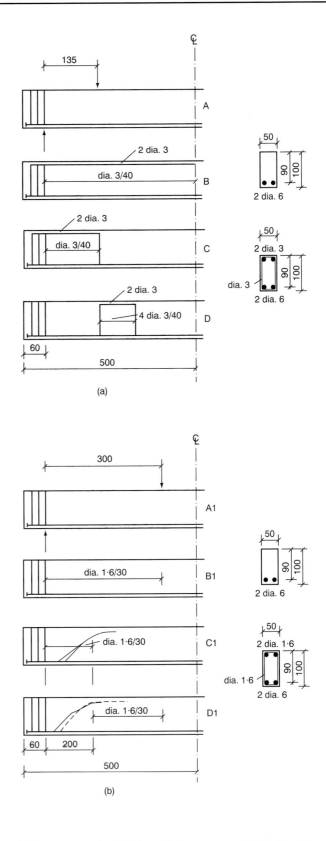

Fig. 2.26. Beams under two-point loading:[2.16] Design details (beams differ in the arrangement of stirrups)

therefore, the shear capacity of the *critical cross-sections* of the portions of the shear spans without transverse reinforcement correspond to a value of the beam's load-carrying capacity significantly smaller than that leading to flexural failure, it would

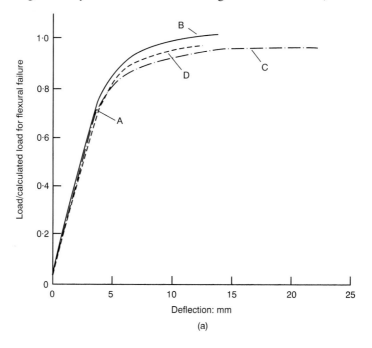

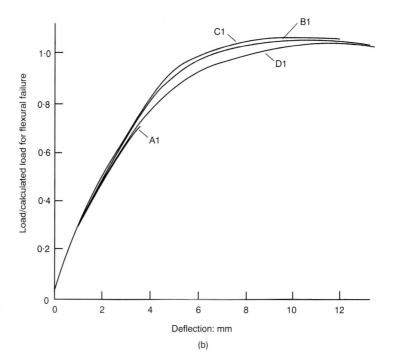

Fig. 2.27. Experimental load–deflection (of the mid cross-section) curves of the beams in Fig. 2.26

be expected that the load-carrying capacity of beams A and D in Fig. 2.26(a), and beams A1, C1, and D1 in Fig. 2.26(b) should correspond to the shear capacity of the beams. Yet, the experimental results depicted in Figs 2.27(a) and 2.27(b) show that, in contrast with beams A and A1 which did indeed fail in 'shear', beam D in Fig. 2.26(a) and beams C1 and D1 in Fig. 2.26(b) exhibited a flexural mode of failure. It may be noted that the load-deflection curves of the above beams are similar to those describing the behaviour of beams B and C (see Fig. 2.27(a)), and beam B1 (see Fig. 2.27(b)) which were designed to current code provisions. From such experimental results, it becomes apparent that the assumption of *shear capacity of critical cross-sections* is not valid.

2.4.1.2. Aggregate interlock

The ductility which characterises beams D and D1 is directly related to the large width of the cracks forming within the tensile zone as the beams approach their ultimate limit state. It is important to note that the width of the inclined crack which formed within the portion of the shear span without shear reinforcement exceeded 1 mm.[2.16] It has been established experimentally that such a crack width precludes *aggregate interlock* even if there were shearing movement of the crack interfaces.[1.12] In this manner, the present tests verify experimentally that there can be no contribution to the shear capacity through *aggregate interlock* at the interfaces of inclined cracks (a conclusion also corroborated by numerical modelling[1.10, 2.12]).

2.4.1.3. Dowel action

There has also been experimental information indicating that *dowel action* cannot contribute to shear capacity. *Dowel action* is effected by the bending and shear stiffnesses of a steel bar,[2.18] and, as a result, it must be affected by the diameter of such bars. A reduction in bar diameter should lead to a considerable reduction of flexural and transverse stiffnesses and, hence, it is realistic to expect a significant reduction in the contribution of *dowel action* to shear capacity. However, a reduction in the diameter of the bars used as longitudinal reinforcement for beams such as beams A and A1 in Figs 2.26(a) and 2.26(b) respectively, in a manner that maintains the total amount of longitudinal reinforcement essentially constant (see Fig. 2.28), was found to have no effect on the shear capacity of beams.[2.18]

2.4.1.4. Truss analogy

From the experimental information presented in the preceding sections, it became apparent that beams D, C1 and D1 did not fail in shear despite the fact that the absence of stirrups within the shear span of beam D in Fig. 2.26(a) and within a large portion of

Fig. 2.28. Reinforcement details for the series of beams under two-point loading (tested for dowel action$^{2.18}$) comprising various bars with the same total cross-sectional area and position of centroid

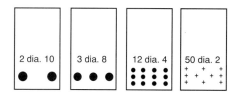

the shear spans of beams C1 and D1 in Fig. 2.26(b) precludes the transformation of the beams into trusses. It would appear, therefore, that it is *not* a necessary condition for a beam with stirrups to behave (at the ultimate limit state) as a truss in order to resist the action of shear forces. Moreover, the experimental results in sections 1.4 to 1.6 indicated that to design structural concrete members (in compliance with current code provisions) such that they behave as trusses at the ultimate limit state does not always safeguard against 'shear' types of failure. In view of the above, it is concluded that there is no justification for the assumption that a beam-like member with both longitudinal and transverse reinforcement behaves as a truss as soon as it suffers inclined cracking.

2.4.2. Contribution of compressive zone to shear capacity

As it has been established from the experimental information presented in the preceding sections that, in the absence of transverse reinforcement, the tensile zone of shear spans cannot contribute to a beam's shear capacity, it can only be concluded that the compressive zone is the *sole* contributor to shear capacity. Such a conclusion is reinforced by the experiments on beams depicted schematically in Fig. 2.29 which led to the data presented in Fig. 2.30.$^{2.19}$ The latter figure shows that restricting the use of stirrups only within the *compressive* zone of beam B in Fig. 2.29 does not essentially alter the mechanical characteristics of the beam in comparison with those of beam A which was designed in compliance with current code provisions. Moreover, a reduction in the spacing of the stirrups within the compressive zone, as indicated in beam C of Fig. 2.29, leads to a considerable improvement of the beam response as regards both strength and stiffness, without affecting ductility. This is because the denser spacing of the stirrups provides effective confinement to the compressive zone, thus increasing the compressive strength of concrete by an amount which maintains the force sustained by the compressive zone, in spite of the reduction of the zone depth caused by the longer extension of the flexural cracks at the higher load attained. (The reduction of the compressive-zone depth increases the lever arm of the internal longitudinal actions (compression sustained by the

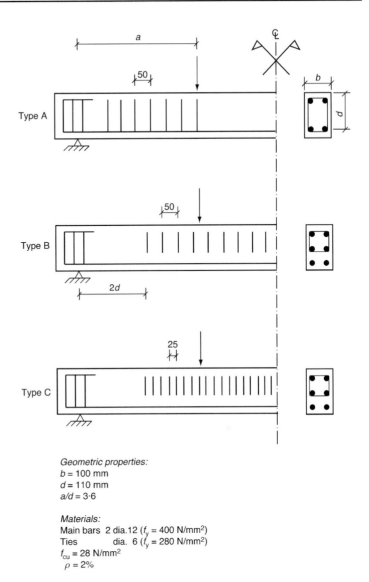

Fig. 2.29. Beams under two-point loading:[2.19] design details (beams differ only in the arrangement and the leg length of the stirrups)

Geometric properties:
$b = 100$ mm
$d = 110$ mm
$a/d = 3 \cdot 6$

Materials:
Main bars 2 dia.12 ($f_y = 400$ N/mm²)
Ties   dia. 6 ($f_y = 280$ N/mm²)
$f_{cu} = 28$ N/mm²
$\rho = 2\%$

compressive zone and tension sustained by longitudinal reinforcement), thus leading to an increase in flexural capacity, without the need for an increase in the tensile force sustained by the longitudinal reinforcement.)

An increase in shear capacity may also result from an increase of the cross-sectional area of the compressive zone of a beam. It is well established that the shear capacity of a beam with a T-section increases by approximately 20% with respect to that of a beam with a rectangular cross-section with characteristics similar to those of the web of the T-beam.[2.4] Since such an increase is insignificant in comparison with the wide scatter of the values of the shear capacity established from tests on a wide range of beams,[1.10, 2.20] current code provisions ignore it and recommend

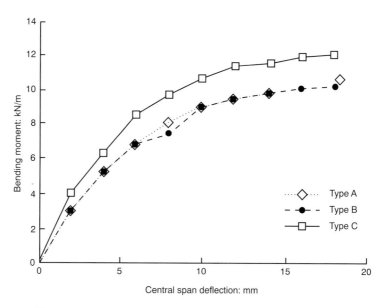

Fig. 2.30. Experimental load–deflection (of the mid cross-section) curves of the beams in Fig. 2.29

Note: Full bending-moment capacity = 9·2 kN/m

that the shear capacity of a T-section should correspond to a rectangular section equal in size to the web of the T-section.[1.3, 1.6]

There are cases, however, for which beams with a T-section exhibit a shear capacity significantly higher than that predicted by current design methods. A typical such case is that of a beam with the geometric characteristics shown in Fig. 2.31 which was found to have a shear capacity more than three times that predicted by current code provisions.[2.21] This experimentally established value correlates closely with the prediction of a semi-empirical expression proposed for the assessment of shear capacity.[2.24] This expression places significant importance, among other parameters, on the *shape* of the section, which is ignored completely by current code provisions. The effect of the shape of the section is related to the 'smoothness' of the flow of the internal stresses which develop within a beam in flexure.[2.23] Full details of the theoretical basis and the predictions of the above expression will be given in Chapter 4.

2.4.3. Shear-failure mechanism

The crack pattern of the beam in Fig. 2.31 is shown in Figs 2.32 and 2.33 for values of the applied load equal to 63 kN and 135kN, respectively. The former value of the applied load is nearly double the value predicted by current codes to cause a 'shear' type of failure, while the latter value is about four times larger than the code prediction. It is interesting to note in the figures that, in spite of the considerable increase of the applied load, the crack patterns differ only in the width of the inclined crack, which attained a value exceeding 3 mm for the case of the higher

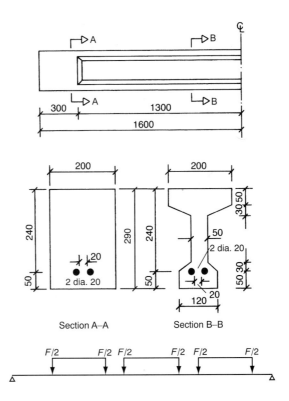

Fig. 2.31. Design details and loading arrangement of T-beams tested under six-point loading[2.21]

load. As for the case of beams D and D1 in Figs 2.26(a) and 2.26(b) respectively, such a crack width precludes *aggregate interlock* along the crack surfaces.[1.12]

However, the main characteristic of the crack pattern — in both cases — is the deep penetration of the inclined crack into the compressive zone which, at the cross-section including the tip of the inclined crack has a depth of merely 10 mm. For the two values of the applied load considered above, the shear force acting at this cross-section attains values of 10·5 kN and 25 kN respectively. As the size of the crack width precludes any contribution to shear capacity other than that of the compressive zone, the mean values of the shear stress corresponding to the above values of the shear force are 5·25 MPa and 12·5 MPa respectively. These values of the shear stresses are indicative of the magnitude of the tensile stresses expected, in accordance with current design methods, to develop within the compressive zone in the region of the tip of the deep inclined crack. As the magnitude of the tensile stresses exceeds by a large margin the tensile strength of concrete ($f_t = 0·1 \times f_c = 0·1 \times 32 = 3·2$ N/mm²), failure should have occurred well before the lower of the values of the applied load considered above was attained.

However, current design methods ignore the existence of a triaxial compressive stress field in the region of the tip of the deepest inclined crack, which is caused by the local volume

REAPPRAISAL OF METHODS FOR STRUCTURAL CONCRETE DESIGN 71

Fig. 2.32. Crack pattern of the beam in Fig. 2.31 under a load level approximately half its load-carrying capacity

dilation of concrete under the large longitudinal compressive stresses which, as discussed in section 2.3.2, inevitably develop at a cross-section with a small depth of the compressive zone. The existence of such a triaxial compressive stress state counteracts the tensile stresses due to the shear forces acting in the same region in the manner schematically described in Fig. 2.34, and, hence, the stress conditions remain compressive in this region, in spite of the presence of exceedingly large shear stresses.

Finally, with a further increase of the applied load, the values of the shear forces increase to a level at which the tensile stresses that they cause cannot be counteracted by the compressive stresses developing by volume dilation. As for the case of the mechanism of flexural failure (see section 2.3.2), failure of the compressive zone is characterised by the development of longitudinal cracking, as indicated in Figs 2.35(a) and 2.35(b) which show the crack patterns at the instant 'shear' types of failure occur. The former figure refers to the beam of Fig. 2.31, subjected to two-point loading (instead of the six-point loading which resulted in the crack patterns in Figs 2.32 and 2.33), while the latter refers to beam D1 in Fig. 2.26(b).

2.4.4. Contribution of transverse reinforcement to shear capacity

As discussed in section 2.4.1.4, from the experimental information presented in the preceding sections, it becomes apparent that it is not essential for a beam with transverse reinforcement to behave (at the ultimate limit state) as a truss in

Fig. 2.33. Failure mode of the beam in Fig. 2.31

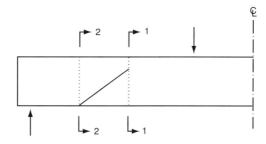

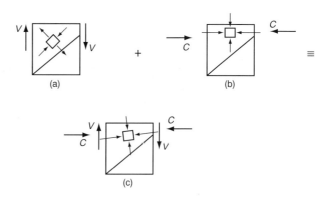

Fig. 2.34. Schematic representation of stress conditions in the region of the tip of a deep inclined crack: (a) due to the shear force; (b) due to the compressive force caused by bending; and (c) due to the combined action of the compressive and shear forces

(a)

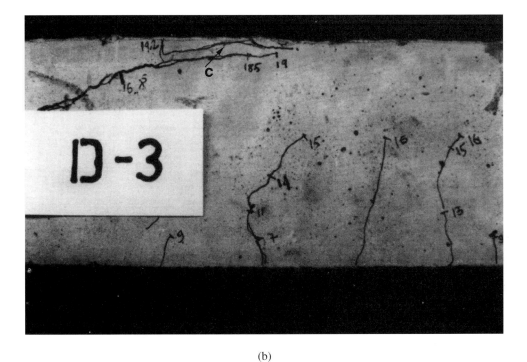

(b)

Fig. 2.35. Horizontal cracking which precedes failure of the compressive zone of the (a) beam in Fig. 2.31 under two-point loading, and (b) beam D1 in Fig. 2.26(b)

order to resist the action of a shear force. The absence of such reinforcement within the shear span of beam D in Fig. 2.26(a), a large part of the shear spans of beams C1 and D1 in Fig. 2.26(b), and within the tensile zone of beams B and C in Fig. 2.29, precludes the transformation of the beams into trusses. Moreover, even for the case of beams designed in compliance with current code provisions, the presence of transverse reinforcement is not sufficient for ensuring *truss action* without the contribution of cracked concrete within the tensile zone.

As discussed in section 2.3.1.1, macrocracking dictates the strain-softening characteristics of concrete. However, in accordance with the experimental information presented in the above section, after the formation of macrocracks, concrete behaviour is characterised by a complete and immediate loss of load-carrying capacity. Hence, in contrast with current design methods, the inability of cracked concrete to allow for the formation of inclined struts precludes the transformation of a beam into a truss (as noted in section 2.4).

However, all experimental information published to date indicates that the provision of transverse reinforcement results in a considerable increase of shear capacity. It is realistic to postulate that the mechanism of the contribution of transverse reinforcement to shear capacity is similar to that of the longitudinal reinforcement to flexural capacity, in that the reinforcement is capable of sustaining the portion of the tensile actions (in the direction of the reinforcement) that cannot be sustained by concrete alone. However, the manner in which transverse reinforcement may be used in an efficient manner is discussed in Chapter 4.

2.5. Conclusions

From the experimental evidence presented in this chapter, it becomes apparent that many of the concepts underlying current design methods are incompatible with fundamental characteristics (such as, for example, stress–strain relations, cracking processes, failure mechanism, etc.) of the behaviour of concrete at the material level. It is a consequence of this incompatibility that assumptions such as 'the uniaxial stress–strain response of concrete in the compressive zone of a 'critical' cross-section in flexure', the contributions of 'aggregate interlock' and 'dowel action', the 'truss analogy', etc., are not borne out by experimental evidence.

The experimental results also indicate that the compressive zone of a beam-like member at its ultimate limit state is subjected to triaxial stress conditions which results in a shear capacity considerably larger than that corresponding to plane-stress conditions. This additional shear capacity compensates for the inability of cracked concrete within the tensile zone to contribute to the shear capacity of the member.

When either the flexural or the shear capacity of a structural concrete beam-like member is exhausted, failure was found, in both cases, to be caused by failure of the compressive zone. The mechanism of such failure is related to the development of transverse tensile stresses which, when the tensile strength of concrete is exceeded, cause longitudinal cracking of the compressive zone which propagates rapidly and leads to collapse.

2.6. References

2.1. Kotsovos M.D. and Newman J.B. Behaviour of concrete under multiaxial stress. *ACI Journal*, 1977, **74**, No. 9, September, 453–456.

2.2. Kotsovos M.D. Fracture of concrete under generalised stress. *Materials & Structures, RILEM*, 1979, **12**, No. 72, November–December, 151–158.

2.3. Kotsovos M.D. and Newman J.B. Fracture mechanics and concrete behaviour. *Magazine of Concrete Research*, 1981, **33**, No. 115, June, 103–112.

2.4. Kong F.K. and Evans R.H. *Reinforced and prestressed concrete*. Van Nostrand Reinhold, Wokingham, 1987, 3rd edn.

2.5. Kotsovos M.D. Effect of testing techniques on the post-ultimate behaviour of concrete in compression. *Materials & Structures, RILEM*, 1983, **16**, No. 91, January–February, 3–12.

2.6. Kotsovos M.D. Deformation and failure of concrete in a structure. *Int. Conf. on Concrete under Multiaxial Conditions, RILEM-CEB-CNRS*, Toulouse, May 1984, **1**, 104–113.

2.7. van Mier J.G.M. Multiaxial strain-softening of concrete. *Materials & Structures, RILEM*, 1986, **19**, No. 111, May–June, 179–200.

2.8. van Mier J.G.M., Shah S.P., Arnaud M., Balayssac J.P., Bascoul A., Choi S., Dasenbrock D., Ferrara G., French C., Gobbi M.E., Kasihaloo B.L., König G., Kotsovos M.D., Labuz J., Lange-Kornbak D., Markeset G., Pavlović M.N., Simsch G., Thienel K-C., Turatsinze A., Ulmer U., van Geel H.J.G.M., van Vliet M.R.A., Zissopoulos D. Strain-softening of concrete in uniaxial compression. *Materials & Structures, RILEM*, 1997, **30**, No. 198, May, 195–209. (Report of the Round Robin Test carried out by RILEM TC 198–SSC: test methods for the strain-softening response of concrete.)

2.9. Kotsovos M.D. A fundamental explanation of the behaviour of reinforced concrete beams in flexure based on the properties of concrete under multiaxial stress. *Materials & Structures, RILEM*, 1982, **15**, No. 90, November–December, 529–537.

2.10. Kotsovos M.D. Concrete — A brittle-fracturing material. *Materials & Structures, RILEM*, 1984, **17**, No. 98, March-April, 107–115.

2.11. Kotsovos M.D. Plain concrete under load — A new interpretation. *Proc. IABSE Colloquium on Advanced Mechanics of Reinforced Concrete*, Delft, June 1981, 143–158.

2.12. Kotsovos M.D. and Pavlović M.N. Non-linear finite-element modelling of concrete structures: Basic analysis, phenomenological insight, and design implications. *Engineering Computations*, 1986, **3**, No. 3, September, 243–250.

2.13. Kotsovos M.D. Consideration of triaxial stress conditions in design: a necessity. *ACI Structural Journal*, 1987, **84**, No. 3, May–June, 266–273.

2.14. British Standards Institution. *Structural use of concrete. Part 2. Code of practice for special circumstances.* BSI, London, 1985, BS 8110.
2.15. Taylor R. and Al-Najmi A.Q.S. The strength of concrete in composite reinforced concrete beams in hogging bending. *Magazine of Concrete Research*, 1980, **32**, No. 112, September, 156–163.
2.16. Kotsovos M.D. *Shear failure of reinforced concrete beams: a reappraisal of current concepts.* Comité Euro-International du Béton, Bulletin d'Information, 1987, No. 178/179, 103–112.
2.17. Kotsovos M.D. Compressive force path concept: basis for reinforced concrete ultimate limit state design. *ACI Structural Journal*, 1988, **85**, No. 1, January–February, 68–75.
2.18. Jelić I., Pavlović M.N. and Kotsovos M.D. A study of dowel action in reinforced concrete beams. *Magazine of Concrete Research*, forthcoming.
2.19. Kuttab A.S. and Haldane D. Detailing for shear with the compressive force path concept. *IABSE Colloquium on Structural Concrete*, Stuttgart, April 1991, IABSE Report 62, 661–666.
2.20. Allen A.H. Reinforced concrete design to CP 110 — Simply explained. *Cement and Concrete Association*, London, 1977.
2.21. Kotsovos M.D., Bobrowski J. and Eibl J. Behaviour of reinforced concrete T-beams in shear. *The Structural Engineer*, 1987, **65B**, No. 1, March 1–10.
2.22. Bobrowski J. and Bardhan-Roy B. K. A method of calculating the ultimate strength of reinforced and prestressed concrete beams in combined flexure and shear. *The Structural Engineer*, 1969, **47**, No. 5, May, 197–209.
2.23. Bobrowski J. *Origins of safety in concrete structures.* University of Surrey, 1982, PhD thesis.

3. The concept of the compressive-force path

3.1. Introduction

This chapter proposes a qualitative description of the behaviour and function of a structural concrete member at its ultimate limit state, together with a description of the mechanism which underlies the transfer of external load from its point of application to the supports of the structural member. This qualitative description, which is compatible with the experimental information presented in Chapter 2, is made by reference to the case of a simply-supported beam, without stirrups, at its ultimate limit state under transverse loading (the effect of axial loading is also considered). Such a structural member was chosen because, not only is there ample experimental information describing its behaviour but, also, the description of how the beam actually functions forms the theory underlying the development of the design methodology proposed in the next chapter. This theory has been termed the 'compressive-force path (CFP) concept' since, as deduced from the description of how the beam functions, the main characteristic of the beam is that both its loading capacity and failure mechanism are related to the region of the member containing the path of the compressive stress resultant which develops within the beam due to bending, just before failure occurs. Experimental information on the validity of the proposed concept is also presented, and it is shown that the concept provides a realistic description of the fundamental causes which dictate the various types of beam behaviour as established by the experimental information available to date. The generalisation of the proposed concept, so as to extend its applicability to any structural configuration and, in particular, to the case of skeletal structures, forms part of the subject presented in the next chapter.

3.2. Proposed function of simply-supported beams

3.2.1. Physical state of beam

Figure 3.1 provides a schematic representation of the crack pattern and the deflected shape (in a magnified form) of a simply-supported beam under transverse loading, just before failure. The figure shows that cracking encompasses a large portion of the beam and comprises both vertical and inclined cracks. The cracks, in most cases, initiate at the bottom face of the beam and, having propagated through the beam web, penetrate deeply into the compressive zone, the crack tip moving closer to the upper face. As will be seen in section 3.3, when the causes of failure are associated with the presence of the deep inclined crack closest to

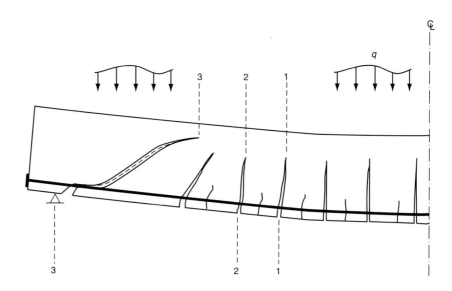

Fig. 3.1. Schematic representation of crack pattern and deformed shape of a simply-supported RC beam under transverse load

the support, this crack not only penetrates into the compressive zone deeper than any other crack, but also extends towards the support along the longitudinal tension bars, destroying the bond between the bars and the surrounding concrete.

It would appear from Fig. 3.1, therefore, that concrete eventually remains uncracked only within a relatively small portion of the beam. This portion includes, on the one hand, the two end regions of the beam which extend to the deep inclined crack forming closest to the supports and, on the other hand, the relatively narrow strip, with varying depth, which forms between the crack tips and the upper face, and connects the above two end regions. As will become apparent in what follows, a characteristic feature of the above narrow strip is its very small depth which, as indicated in the figure, is, in localised regions (and, in particular, in the region including the tip of the deepest inclined crack), a very small percentage of the total beam depth.

It should be noted that the presence of an external load acting on the end faces of the beam, in the axial direction, may have the following two effects on the physical state of the beam depicted in Fig. 3.1.

>   (a) The depth of the horizontal uncracked zone of the beam may increase or decrease (leading to a corresponding reduction or increase in the length of the flexural cracks), depending on whether the axial force is compressive or tensile respectively.
>   (b) The presence of the axial force may prevent the formation of any deep inclined crack.

In all other respects, the physical state of the beam should be qualitatively similar to that depicted in Fig. 3.1.

3.2.2. Load transfer to supports

In spite of the extensive cracking, the beam at its ultimate limit state is capable of fulfilling its purpose, i.e. transferring the applied load to the supports. The mechanism through which this transfer is effected can only be a form of 'beam action' adjusted so as to allow for the particular characteristics of reinforced-concrete members.

In any beam cross-section (in which the presence of an external axial force is ignored for purposes of simplicity), internal actions may be resolved into axial and transverse components. In particular, for the case of a cross-section including a deep flexural crack (such as, for example, cross-section 2–2 in Fig. 3.1), the axial internal actions are such that their combined action is equivalent to the bending moment which develops in this cross-section as a result of the external load, while the shear force is equivalent to the resultant of the external transverse forces acting on the beam portion to the left of the cross-section in question (see Fig. 3.2).

The relationship between the internal axial and shear forces may be derived by considering the equilibrium conditions of an element of the beam between two cross-sections including consecutive flexural cracks such as, for example, the element between cross-sections 1–1 and 2–2 in Fig. 3.1, which is also illustrated in isolation as a free body in Fig. 3.3. The action of the couple arising from the shear forces that develop at the two end cross-sections of this element equilibrates the change in the bending moment between these two cross-sections. This change of bending moment is predominantly due to the change in the magnitude of the axial internal actions, i.e. the compressive force sustained by concrete and the — numerically equal to it (for purposes of equilibrium) — tensile force sustained by the longitudinal steel bars (see Fig. 3.3(b)).

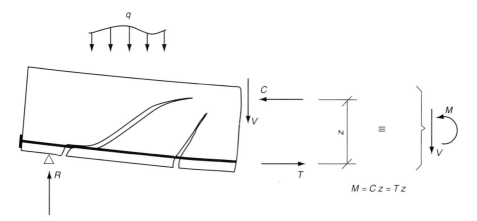

$M = Cz = Tz$

Fig. 3.2. Internal actions equivalent to the bending moment and shear force acting at a cross-section including a crack

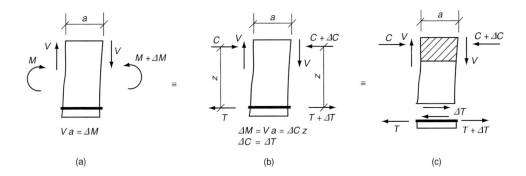

Fig. 3.3. Equilibrium of portion of beam (in Fig. 3.1) between two cross-sections including consecutive cracks

A necessary prerequisite for the change in magnitude of the above longitudinal internal actions is the existence of bond between concrete and steel, through which a portion (ΔT) of the tensile force acting on the steel bars is transferred to the concrete (see Fig. 3.3(c)). It should be noted that force ΔT is the *only* action developing on any of the concrete strips between consecutive flexural or inclined cracks, since the experimental evidence presented in the preceding chapter precludes the development of any significant forces at the crack surfaces due to 'aggregate interlock,' while 'dowel action', even if it were to develop, is negligible.

A concrete strip such as the above may be considered to function as a 'cantilever' fixed on to the compressive zone of the beam and subjected to the action ΔT transmitted from steel to concrete through bond (see Fig. 3.3(c)).[3.1] The bending moment that develops at the cantilever base, owing to the above force, balances the action arising from the couple of the shear forces which act in the compressive zone of the beam. In fact, the above equilibrium condition essentially describes the mechanism through which the external load, in the form of shear forces, is transferred throughout the length of the span within which bond develops between concrete and steel (see Fig. 3.4(a)).

However, as discussed in section 3.2.1, the existence of the deep inclined crack near the beam support causes bond failure, the latter extending between the intersection of the inclined crack with the longitudinal steel bars and the support, and thus the external load cannot be transferred by 'cantilever bending' beyond the cross-section which includes the tip of the inclined crack. The mechanism through which the external load is transferred from the above section to the support becomes apparent by considering the equilibrium conditions of the end portion of the beam which encompasses the region enclosed by the end, upper, and lower faces of the beam, the inclined crack closest to the support, and the cross-section through the tip of this crack. This portion is isolated from the beam and represented schematically by the free body illustrated in Fig. 3.4(b).

THE CONCEPT OF THE COMPRESSIVE-FORCE PATH

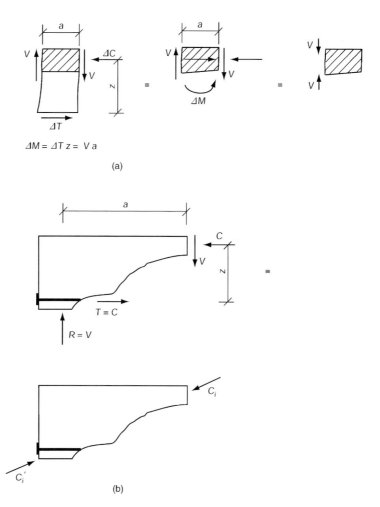

Fig. 3.4. Mechanisms of external-load transfer to the supports: (a) cantilever action, and (b) change in direction of compressive force

Owing to the destruction of the bond between concrete and the longitudinal bars, the tensile force sustained by the reinforcement is transmitted unchanged from the right-hand side of the free body to the region of the support where it combines with the reaction to yield the inclined compressive force C'_i (see Fig. 3.4(b)). Similarly, the shear and compressive-axial forces acting on the upper part of the right-hand side end of the free body combine to form the inclined compressive force C_i (see also Fig. 3.4(b)) which, for equilibrium purposes, must fulfil the condition $C_i = C'_i$. This condition indicates that it is through the development of the inclined compressive force C_i that the external load is transferred from the right-hand side of the free body to the support. In fact, the development of the above inclined force essentially represents a change in the direction of the path of the near-horizontal (within the middle portion of the beam's) compressive stress resultant which develops on account of the bending of the beam, with the

3.2.3. Effect of cracking on internal actions

change in path direction occurring in the region of the tip of the inclined crack closest to the support.

An indication of the internal state of stress and the magnitude of the stresses which develop in cracked concrete may also be obtained by considering the forces acting on the beam element illustrated in Fig. 3.3. This element lies between two cross-sections (1–1 and 2–2 in Fig. 3.1) which include consecutive cracks and, as discussed in the preceding section, the only action exerted on the tensile zone of the element is the portion of the tensile force (ΔT in Fig. 3.3(c)) which is transferred from the longitudinal steel bars to concrete through bond. In fact, the crack surfaces which form boundaries to this element remain stress-free since, in accordance with the experimental evidence presented in the preceding chapter, the cracking mechanism of the beam precludes the development of both 'aggregate interlock' and 'dowel action' which are the most likely mechanisms that could allow for the development of forces at the crack faces.

As discussed in the preceding section, therefore, the portion of this element between the cracks acts as a plain-concrete cantilever (fixed to the compressive zone of the beam) which undergoes bending as a result of the tensile force ΔT transmitted from the steel bars to concrete through bond (see Fig. 3.5(a)). The state of stress which is compatible with 'cantilever bending' results from the development of, on the one hand, a shear force constant throughout the cantilever length and equal to ΔT (see Fig. 3.5(c)), and, on the other hand, a bending moment, the magnitude of which increases with the distance from the free end of the cantilever, attaining its maximum value at the cross-section (3–3 in Fig. 3.5(a)) which coincides with the fixed end.

Figures 3.5(b) and 3.5(c) show the distributions of the normal (σ') and shear (τ') stresses and the corresponding stress resultants ($T' = C'$, V'), whose combined action ($T'z' = C'z$, V') is, for purposes of equilibrium, equivalent to that of the bending moment ($\Delta M = \Delta Tz = Va$ (see Fig. 3.5(a)) and shear force ($V' = \Delta T$) acting on the cross-section (3–3 in Fig. 3.5(a)) coinciding with the fixed end of the cantilever. Since the cantilever consists of plain concrete, the (numerically) maximum value of the stresses developing at the above cross-section cannot exceed the strength (f_t) of concrete in tension (as indicated in Figs 3.5(d) and 3.5(e) which depict the stress conditions at two typical elements E1 and E2 (of cross-section 3–3 in Fig. 3.5(a)) in pure tension and pure shear respectively). (For such stress values, concrete behaviour is essentially linear and, hence, the shape of the stress distributions assumed in Figs 3.5(b) and 3.5(c) is that predicted by the simplified beam theory.)

THE CONCEPT OF THE COMPRESSIVE-FORCE PATH

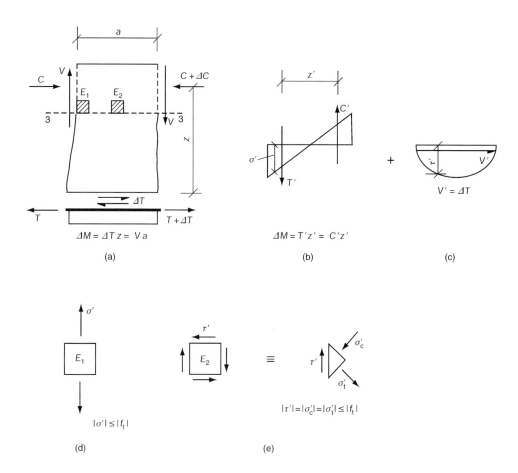

Fig. 3.5. State of stress at the base of the 'cantilever' under the action of the bond force

As the fixed end of the cantilever essentially coincides with the interface between the uncracked (compressive) and cracked (tensile) zones of the beam, the uncracked zone is also subjected to the internal stresses and stress resultants acting at this interface, as indicated in Figs 3.5(b) and 3.5(c). However, the main actions that develop within uncracked concrete are those indicated in Fig. 3.2 which, together with the tensile force sustained by the longitudinal reinforcement, resist the combined action of the bending moment and shear force caused by the applied load. Figure 3.2 indicates that uncracked concrete (i.e. the compressive zone of the beam cross-section 2–2 in Fig. 3.1) is subjected not only to the axial compressive force C (due to the bending moment) but also to the *total* shear force acting at the beam cross-section (since, as discussed earlier, cracked concrete cannot contribute to the shear resistance of the beam). Here, it should be recalled that, although the magnitude of the nominal shear stress (i.e. the ratio of the shear force to the area of the compressive zone of the cross-section) exceeds (in regions where the depth of the uncracked concrete is small) the concrete shear capacity (as defined in current codes) by a large margin, the mechanism of shear resistance described in section 2.4.3 (see also Fig. 2.34) enables uncracked concrete to sustain the applied shear force. In compliance with this mechanism, the presence of triaxial stress conditions (in localised regions of the compressive zone where the depth is small) delays the development of tensile stresses (caused by the shear force); therefore, the value of the shear force required to cause failure of the compressive zone becomes significantly larger than that expected to cause failure in compliance with the concepts underlying current design methods.

It should also be noted that, in accordance with the experimental data presented in section 2.3.1.3, the compressive zone of the element illustrated in Fig. 3.5 is subjected to large axial stresses which, owing to the triaxiality of the stress conditions at the ultimate limit state of the beam, may be as large as twice the uniaxial compressive strength (f_c) of concrete.

It would appear from the above outline of the stress conditions in a typical RC beam, therefore, that, while the magnitude of the stresses that develop within cracked concrete cannot exceed a value of the order of the tensile strength of concrete (i.e. a value of approximately 5–10% of the strength of concrete in uniaxial compression), the magnitude of the stresses that develop within uncracked concrete can be of the order of the uniaxial compressive strength of concrete or even larger than it by a factor which, in localised regions, may be as large as 2.

3.2.4. Contribution of uncracked and cracked concrete to the beam's load-carrying capacity

As discussed in the preceding section, uncracked concrete sustains not only the *total* axial compressive force that develops within the beam on account of bending, but also the *total* shear force, the largest portion of which current codes assume to be resisted by cracked concrete through 'aggregate interlock' and 'dowel action.' As a result, the contribution of uncracked concrete essentially represents the *total* contribution of concrete to the load-carrying capacity of the beam.

In contrast, cracked concrete, through the formation of 'plain-concrete cantilevers' between consecutive flexural and/or inclined cracks, provides a mechanism which allows it to make a significant contribution to the transfer of the external load, through the uncracked portion of the beam, from its points of application to the supports. As described in section 3.2.2, this mechanism involves the development of bending moments at the fixed ends of the cantilevers (interface between uncracked and cracked concrete) which balances the actions arising from the shear forces acting at beam cross-sections, including flexural or inclined cracks (as indicated in Fig. 3.4(a)). As described also in section 3.2.2, the development of the bending moments is attributable to the forces ΔT (see Fig. 3.3) which are transferred from steel to concrete (in the free-end region of the cantilever) through bond.

3.2.5. Causes of failure

Figure 3.6 shows a schematic representation of the uncracked portion of the beam as a free body under the action of the external load, applied at its top face, and the action of the internal forces developing along the cut which separates the uncracked portion from the remainder of the beam. The figure also provides an indication of the locations where tensile stresses are likely to develop within the uncracked portion.

As discussed in the preceding sections, the uncracked portion of the beam, through which the applied load is transferred to the supports, encloses the path of the compressive stress resultant which develops within the compressive zone due to the bending of the beam. As discussed in section 3.2.2, this transfer requires, on the one hand, the contribution of the cracked portion of the beam through 'cantilever bending' (the latter causes the internal actions which develop at the interface between the uncracked and cracked regions [see Figs 3.5(b) and (c)]) and, on the other hand, the change in the path direction (see Fig. 3.4(b)) which occurs at the locations where the middle horizontal narrow strip joins the end blocks of the uncracked portion of the beam (see Fig. 3.6).

From the schematic representation of the distribution of the compressive stresses (σ_c) within the end region of the beam shown in Fig. 3.6, it becomes apparent that only a diagonal strip of this region, which forms essentially an extension of the compressive zone, is utilised for the transfer of the applied load to the support. With regard to the remaining portion of this end region, its lower part provides anchorage space for the longitudinal reinforcement, while the upper part remains essentially 'structurally' inert, in the sense that it does not make

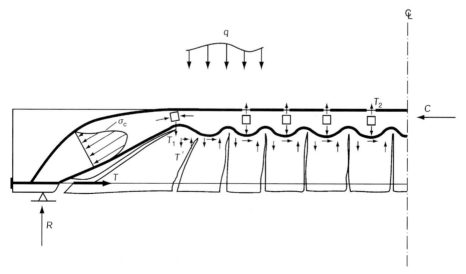

Fig. 3.6. *Crack-free portion of simply-supported beam under the action of external load and internal actions at the interface between cracked and crack-free portions of the beam*

any significant contribution to the transfer of the applied load to the supports, in spite of the development of parasitic stresses of small magnitude and random orientation.

In addition to its contribution to the transfer of the applied load to the supports through bond-induced cantilever action, the presence of the cracked portion of the beam effects the interaction between uncracked concrete and the longitudinal reinforcement, while, at the same time, it maintains the relative position of the above two components of the beam essentially unchanged throughout the loading history of the beam.

Having established in the preceding section that the uncracked portion of the beam is the sole concrete contributor to the load-carrying capacity of the member (the latter being also dependent on the strength of the longitudinal reinforcement), it is essential to identify the causes of beam failure. If it is assumed that the beam is designed so that it does not suffer any loss of its load-carrying capacity as a result of failure of the longitudinal steel bars, then the causes of failure should be sought in the portion of the beam which comprises uncracked concrete, since cracked concrete could be viewed as concrete already failed.

On the basis of the experimental data presented in the preceding chapter, concrete always fails in tension. As a result, the search for the causes of failure of the portion of the beam comprising uncracked concrete only must be focused on the identification of regions of this portion where tensile forces are likely to develop. Such regions may be the following.

(a) *Regions of change in the direction of the path of the compressive stress resultant*. A tensile stress resultant (T_1 in Fig. 3.6) may develop at the location where the path changes direction as a response to the action of the vertical component of the inclined compressive stress resultant, developing within the end block of the uncracked portion of the beam, which tends to separate the upper part of the compressive zone from the remainder of the beam by splitting near-horizontally this zone in the region of the change in the path direction. (The change in direction of the stress trajectory necessitates, for equilibrium purposes, a (nearly-vertical) orthogonal force bisecting the angle between the two stress directions.)

(b) *Interface between uncracked and cracked concrete*. As indicated in Fig. 3.6, tensile actions (in the sense that they pull the 'cracked' regions away from the 'uncracked' ones) develop, as described in section 3.2.3, at the above interface due to 'cantilever bending' in the cracked region of the beam (see Fig. 3.5(b)). Since, as deduced from the expression $T'z' = Va$ in Fig. 3.5, T' is proportional to V, an indication of the variation of the magnitude of T'

within the beam span may be obtained from the shear force. From the latter's diagram, it can be seen that the most likely tensile action to cause failure (i.e. σ' to exceed f_t) is that (T' in Fig. 3.6) which develops in the region of the tensile action T_1 where the compressive-force path (to which the uncracked portion of the beam forms an envelope) changes direction. Failure in this region may occur not only because the tensile action in this region attains the largest value (as indicated by the shear-force diagrams outside the region of the uncracked beam end (in the latter, load transfer does not occur through cantilever bending) of most types of loading condition considered in practice), but also because the inclined crack in this region has the most favourable orientation for crack extension (as opposed to more central portions of the shear span, where the existing (mainly flexural) cracks are near-normal to the cracks caused by T'). (The tangent to the shape of the inclined crack at the crack tip coincides essentially with the orientation of the principal compressive stress which defines the direction of crack extension.)

(c) *Regions adjacent to those including cross-sections with deep flexural or inclined cracks.* Volume dilation of concrete in the compressive zone of regions including cross-sections with deep flexural or inclined cracks induce transverse tensile actions T_2 in the adjacent regions. (A full description of this mechanism for the development of such transverse actions is given in section 2.3.) Four such possible locations are illustrated generally in Fig. 3.6.

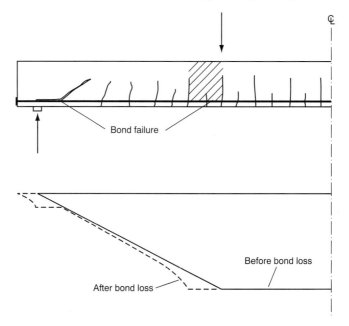

Fig. 3.7. Effect of bond loss on tensile force sustained by longitudinal reinforcement (note that the shape of the variation of the tensile force before bond loss occurs is similar to that of the bending-moment diagram)

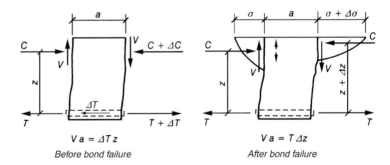

Fig. 3.8.
Redistribution of internal actions in the compressive zone due to loss of bond between concrete and longitudinal steel bars

(d) *Regions of applied point loads.* These regions usually include cross-sections within the shear span where the applied bending moment is large (see Fig. 3.7). At the ultimate limit state of the beam, it is likely for bond failure to occur in the tensile zone of such regions (see Fig. 3.8). From the figure, it can be seen that the loss of bond results in an extension of the right-hand side flexural crack sufficient to cause an increase Δz of the lever arm such that $T\Delta z = Va$ (thus preserving moment equilibrium which would otherwise be lost because of the elimination of ΔT as a result of the bond destruction). The extension of the flexural crack reduces the depth of the neutral axis and this increases locally the intensity of the compressive stress field. In turn, this increase in the stress intensity should give rise to tensile actions in the manner previously described in item *(c)* above. (Therefore, one can conclude that bond failure is likely to occur either near the support (due to the propogation of the inclined crack towards the support along the interface between concrete and longitudinal reinforcement) or at locations of large bending moment (and non-zero shear) because of the tensile yielding of the bars.)

3.3. Validity of the proposed structural functioning of simply-supported beams

The description of the functioning of the simply-supported beam proposed in the preceding sections contrasts with current views with regard to the following points.

(a) Uncracked concrete in compression (through which the applied load is transferred to the supports) and the longitudinal main-steel bars in tension are essentially the sole contributors to the load-carrying capacity of the beam, with cracked concrete contributing mainly to the transfer of applied load to the supports through 'cantilever bending'.

(b) Failure of the beam is caused by the development of tensile stresses within the previously uncracked concrete, which act transversely to the longitudinal compression that develops as a result of the bending of the beam.

THE CONCEPT OF THE COMPRESSIVE-FORCE PATH

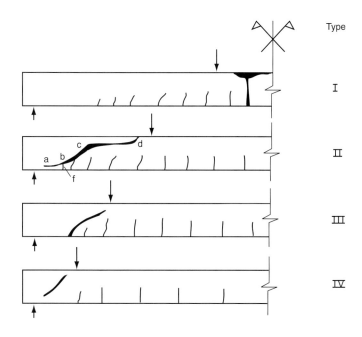

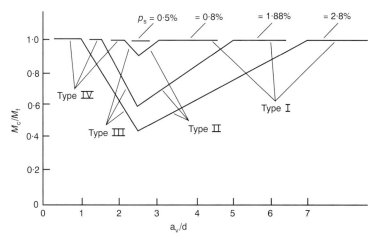

Fig. 3.9.
Characteristic types of behaviour of a simply-supported beam at its ultimate limit state: (a) mode of failure and (b) relation between bending moment corresponding to load-carrying capacity and shear span for various percentages of longitudinal reinforcement

The adoption of point (a) above is fully justified as this premise is compatible with the experimental data presented in the preceding chapter. However, point (b) remains to be proved as compatible with experimental data available to date on beam behaviour at the ultimate limit state. Such data have been summarized in Fig. 3.9, which provides a schematic representation of the variation of the load-carrying capacity of a simply-supported reinforced-concrete beam, without stirrups, under two-point loading with the shear span-to-depth (a_v/d) ratio for various percentages of the longitudinal reinforcement, with the beam's load-carrying capacity being expressed in the form of the bending moment at the mid cross-section. (This ratio M_c/M_f reflects the actual capacity of the beam relative to its full flexural capacity.) From this form of representation of the data (first introduced by Kani[3.1]), it becomes apparent that the behaviour of the above beam, at its ultimate limit state, may be divided into four types of regimes associated with the value of a_v/d.

Type I behaviour   Type I behaviour corresponds to relatively large values of a_v/d (usually larger than 5) and is characterised by a flexural mode of failure. The causes of such a mode of failure are fully described in section 2.3 of the preceding chapter and they have already been incorporated into the proposed qualitative description of beam behaviour (see item *(c)* in section 3.2.5).

Type II behaviour   Type II behaviour corresponds to values of a_v/d between approximately 2 and 5, and is characterised by a brittle mode of failure which is usually associated with the formation of a deep inclined crack within the shear span of the beam. (Brittle failure may also occur owing to near-horizontal splitting of the compressive zone which occurs independently from any web cracking in the region of the point load, as discussed later in this section. Immediately after its formation, the inclined crack (which, for values of a_v/d closer to 2 rather than 5, is essentially an extension of the flexural crack (marked with *f* in Fig. 3.9) closest to the support) extends near-horizontally (branch *c–d* in Fig. 3.9) within the compressive zone towards the point load in an unstable manner, leading to an immediate and total loss of load-carrying capacity of the beam. (This inclined crack may also extend towards the support along the interface between concrete and the steel bars (branch *a–b* in Fig. 3.9), destroying the bond between the two materials, but such an extension may be prevented from leading to failure of the beam by proper anchoring of the steel bars.)

The causes of such a mode of failure are described by items *(a)* and *(b)* in section 3.2.5. These are associated with the development of tensile actions in the region where the path of the compressive force (owing to the bending of the beam) changes direction. Such tensile actions, as discussed in section 3.2.5, may cause splitting of the compressive zone which leads to total loss of the beam load-carrying capacity. Details of the manner in which the above failure process initiates is illustrated in Figs 2.35(a) and 2.35(b) which show that, between the upper face of the beam and the inclined crack closest to the support, in the region of the crack tip, an isolated (deepest) crack (marked with c in the two figures) forms as soon as the tensile strength of concrete is exhausted. (The extension of this crack was prevented by the instantaneous unloading of the beam as soon as the crack appeared. Maintaining the load constant leads to the failure

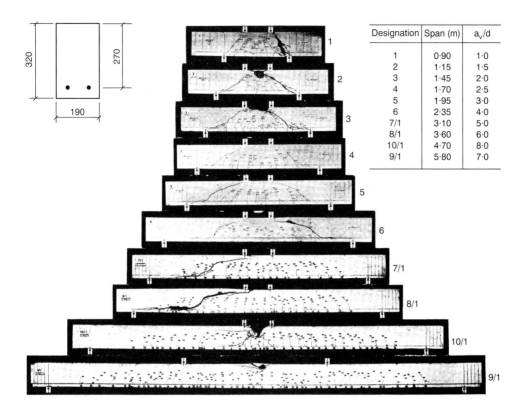

Designation	Span (m)	a_v/d
1	0·90	1·0
2	1·15	1·5
3	1·45	2·0
4	1·70	2·5
5	1·95	3·0
6	2·35	4·0
7/1	3·10	5·0
8/1	3·60	6·0
10/1	4·70	8·0
9/1	5·80	7·0

Fig. 3.10. Modes of failure of beams with various shear spans under two-point loading[3.2]

process described above, which gives the (misguided) impression that failure is caused by the extension of the inclined crack, as usually depicted for type II behaviour (as in Fig. 3.9).) It becomes apparent from the above, therefore, that in order to prevent this type of failure the location at which the compressive force changes direction must be known *a priori*.

Figures 3.10 and 3.11 illustrate the crack patterns of two groups of beams at failure tested under two-point and uniformly-distributed loading respectively.[3.2] As discussed in section 3.2.2, the change in the direction of the path of the compressive force which develops due to the bending of the beam occurs in the region of the tip of the inclined crack forming closest to the support. For the beams under two-point loading with values of a_v/d between 2 and 5 (beams 4 to 6 in Fig. 3.10), as well as for the beams under uniformly-distributed loading with a normalised (with respect to the beam depth) span (L/d) greater than 8 (beams 13 to 17 in Fig. 3.11), the tip of the above crack lies at a distance approximately equal to twice the beam depth ($2d$) from the support. It should be expected, therefore, that the provision of sufficient reinforcement at a distance of $2d$ from the supports should prevent beam failure associated with the causes of failure described by items *(a)* and *(b)* in section 3.2.5. In fact, placing

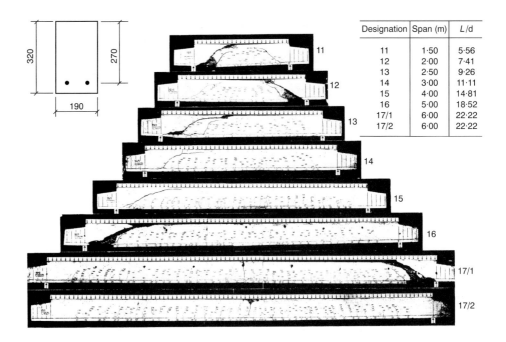

Fig. 3.11. Modes of failure of beams with various spans under uniformly-distributed loading$^{3.2}$

such reinforcement in beam D1 shown in Fig. 2.26(b) allowed the beam to develop its flexural capacity in contrast to the predictions of current methods used for the design of reinforced-concrete structures.

As discussed earlier, beams characterised by type II behaviour may also fail as a result of horizontal splitting of the compressive zone which occurs independently from any web crack. Such splitting may be due to the development of tensile stresses within the compressive zone associated with the loss of bond between concrete and flexural steel as described by item (d) in section 3.2.5. Loss of bond may have been the cause of failure of beams 7 and 8 in Fig. 3.10 whose mode of failure is also characterised by the presence of an inclined crack which formed closer to the load point rather than the support. Although bond failure appears to have occurred in the region between the deep inclined crack and the crack adjacent to it (as one moves away from the support), it cannot be deduced from the modes of failure indicated in the figure that the failure process was that predicted by item (d) in section 3.2.5. Additional information regarding this type of failure will be provided in Chapter 4 as part of the verification study of the new design methodology proposed in that chapter.

Type III behaviour Type III behaviour corresponds to values of a_v/d between approximately 1 and 2 and, as for type II behaviour, is characterised by brittle failure. Such failure is associated with the development of an inclined crack within the shear span of the beam which, as indicated in Fig. 3.9, and in contrast with the inclined crack characterising type II behaviour, forms independently from any pre-existing flexural or inclined crack. Moreover, unlike the inclined crack which characterises type II behaviour, the formation of the inclined crack for type III behaviour does not lead to immediate failure: instead, the applied load must be increased further in order to cause failure of the beam.

The main characteristic of this type of behaviour is that the beam fails outside the shear span. As indicated by the mode of failure of beams 2 and 3 in Fig. 3.10, the extension of the inclined crack, which forms within the shear span, deviates from the region of the applied load, where the strength of concrete is higher owing to the triaxial compressive stress conditions which develop in this region,[3.3, 3.4] and penetrates deeply into the compressive zone of the 'flexure' span of the beam causing failure of the type described by item (c) in section 3.2.5, i.e. the volume dilation of concrete in the compressive zone of the cross-section through the tip of the inclined crack causes transverse tensile stresses in the adjacent regions leading to splitting of the compressive zone and failure of the member, before the flexural capacity is attained.

The above explanation of the causes of failure is compatible with the experimental data obtained from the tests on the beams of type D shown in Fig. 2.26(a). From this figure, it appears that the placing of links only within the 'flexure' span of beams with $a_v/d = 1\cdot6$ delays the formation of horizontal cracks in the region of the point load sufficiently for the beam to exhaust its flexural capacity first. It is also of practical interest, as will be seen in the next chapter, to note that for the case of point loading the change in the direction of the path of the compressive stress resultant occurs in the cross-section through the point load (see beams 2 and 3 in Fig. 3.10), whereas for the case of uniformly-distributed loading this change in path direction occurs at a distance from the support approximately equal to a quarter of the beam span (see beams 11 and 12 in Fig. 3.11).

Type IV behaviour   Type IV behaviour corresponds to values of a_v/d smaller than 1 and is characterised by two possible modes of failure:[3.5] (a) a ductile mode of failure, for the case of failure within the middle narrow strip of the uncracked portion of the beam; and (b) a brittle mode of failure, for the case of failure of the end blocks of the uncracked portion of the beam in the region of the support. As will be seen in Chapter 4, the mode of failure is generally dictated by the size of the beam width, the larger sizes being more likely to lead to a ductile, rather than a brittle, type of failure. It should be noted, however, that in both cases there is no significant, if any, difference in load-carrying capacity. The mechanism of failure described by item *(c)* in section 3.2.5 provides a satisfactory description of the causes of failure which characterises the present type of behaviour.

3.4. Conclusions   The present chapter summarises the experimental data presented in Chapter 2 in a manner that reveals the fundamental characteristics which underlie the behaviour of a simply-supported reinforced-concrete beam, without stirrups, at its ultimate limit state under static monotonic transverse loading. The main conclusions drawn from this summary are as follows.

1. The beam comprises the following:

    (a) an uncracked portion consisting of the two end-regions of the beam which, extending to the (usually) inclined crack forming closest to the support and the cross-section through the tip of this crack, are connected by a narrow strip of varying depth forming between the upper face and the tips of flexural and inclined cracks which initiate at the bottom face and extend towards the upper face of the beam

    (b) a cracked portion consisting of 'plain-concrete cantilevers' which, forming between successive flexural and inclined cracks, are fixed at the narrow zone of the uncracked portion

    (c) the longitudinal reinforcement, penetrating the beam throughout its span at a relatively short distance from the tensile face, fully bonded to concrete at least in the region of the support where it is properly anchored.

2. The uncracked portion encloses the path of the compressive stress resultant (due to the bending of the beam), with a near-horizontal orientation within the middle narrow strip of the uncracked portion, changing in the region of the tip of the inclined crack closest to the support and becoming diagonal within the end regions. The location of the change in the path direction appears to depend on parameters such as, for example, the shear span-to-depth-ratio for the case of point loading, and the span-to-depth ratio for the case of uniformly-distributed loading.

3. The uncracked portion of the beam is not only the sole concrete contributor to the load-carrying capacity of the member but also transfers the applied load to the supports; the cracked portion of the beam contributes to this transfer through cantilever bending.

4. Failure appears to be associated with the development of transverse tensile stresses within the uncracked portion of the beam. The causes for the development of such stresses vary and appear to depend on the value of the parameters referred to in conclusion 2 above.

3.5. References

3.1. Kani G.N.J. The riddle of shear failure and its solution. *ACI Journal*, 1964, **61**, No. 28, April, 441–467.

3.2. Leonhardt F. and Walther R. *The Stuttgart shear tests, 1961. Contributions to the treatment of the problems of shear in reinforced concrete construction.* (A translation (made by C.V. Amerongen) of the articles that appeared in *Beton- und Stahlbetonbau*, 1961, **56**, No. 12 and 1962, **57**, Nos 2–3, 7–8.) Translation No. 111, C&CA, London, 1964.

3.3. Kotsovos M.D. and Newman J.B. Effect of boundary conditions upon the behaviour of concrete under concentrations of load. *Magazine of Concrete Research*, 1981, **33**, No. 116, September, 161–170.

3.4. Kotsovos M.D. An analytical investigation of the behaviour of concrete under concentrations of load. *Materials & Structures, RILEM*, 1981, **14**, No. 83, September–October, 341–348.

3.5. Kotsovos M.D. Design of reinforced concrete deep beams. *The Structural Engineer*, 1988, **66**, No. 2, January, 18–32.

4. Design methodology

4.1. Introduction

In this chapter, the qualitative description of beam behaviour presented in Chapter 3 is transformed into a new methodology suitable for the design of concrete structures. This earlier qualitative description is now condensed into a physical model of a simply-supported beam with behavioural characteristics (such as, for example, crack pattern, internal actions, mechanism of external load transfer to the supports, failure mechanism, etc.) similar to those of a real beam at its ultimate limit state. The physical model forms the basis of the proposed design methodology which is initially developed so as to be used for the design of simply-supported reinforced concrete beams. Its development is then complemented so as to extend its use to the case of prestressed concrete beams, and, finally, it is completed so as to become capable of providing design solutions for any type of skeletal structural-concrete configuration comprising beam-like elements.

4.2. Simply-supported reinforced concrete beam

4.2.1. Physical model

Figure 4.1(a) depicts the physical model of a simply-supported beam, the qualitative characteristic features of which were described in detail in the preceding chapter. Figure 4.1(b) provides a schematic representation of the effect that the presence of external axial load has on this physical model. As indicated in Fig. 4.1(a), the beam, in all cases, is modelled as a 'comb-like' structure with 'teeth' fixed on to the horizontal element of a 'frame' with inclined legs. The 'frame' and the 'teeth' also interact through a horizontal 'tie' which is fully bonded to the 'teeth', near their bottom face, and anchored at the bottom ends of the 'frame' legs. A comparison between the proposed model and the beam of Fig. 3.1 indicates the following.

(a) The 'frame' provides a simplified representation of the uncracked region of the beam which encloses the path of the compressive stress resultant that develops due to bending.
(b) The 'tie' represents the flexural reinforcement.
(c) The 'teeth' of the 'comb-like' model represent the plain-concrete cantilevers which form between successive flexural or inclined cracks within the tensile cracked zone of the beam.

As concluded in the preceding chapter, the load-carrying capacity of the beam is provided by the combined action of the

DESIGN METHODOLOGY

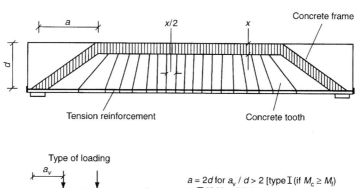

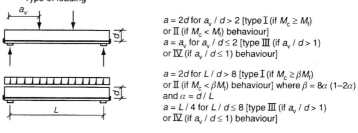

$a = 2d$ for $a_v / d > 2$ [type I (if $M_c \geq M_t$)
or II (if $M_c < M_t$) behaviour]
$a = a_v$ for $a_v / d \leq 2$ [type III (if $a_v / d > 1$)
or IV (if $a_v / d \leq 1$) behaviour]

$a = 2d$ for $L / d > 8$ [type I (if $M_c \geq \beta M_t$)
or II (if $M_c < \beta M_t$) behaviour] where $\beta = 8\alpha (1-2\alpha)$
and $\alpha = d / L$
$a = L / 4$ for $L / d \leq 8$ [type III (if $a_v / d > 1$)
or IV (if $a_v / d \leq 1$) behaviour]

(a)

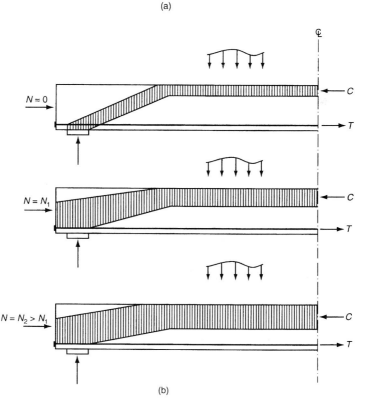

(b)

Fig. 4.1. (a) Physical model of simply-supported RC beam under transverse loading and (b) effect of axial load on physical model in (a)

uncracked concrete and the flexural reinforcement, i.e. the 'frame' and the 'tie' of the proposed model, with uncracked concrete, i.e. the 'frame', also transferring the applied load to the supports, while cracked concrete in the tensile zone, i.e. the 'teeth' of the 'comb', provides the (bond-based) mechanism through which the transfer loop is completed.

4.2.2. Failure criterion

In order to implement in practical design the physical model presented in the preceding section, it is essential to complement it with a failure criterion capable of predicting both load-carrying capacity and mode of failure. Such a failure criterion must be compatible with experimental information such as that summarised, in a pictorial form, in Fig. 3.9. The figure includes a graphical description of beam load-carrying capacity, together with schematic representations of the modes of failure characterising the four distinct types of behaviour indicated in the figure.

It is relatively straightforward to predict the load-carrying capacity for the case of type I behaviour since, as indicated in Fig. 3.9, the ultimate capacity corresponds to the flexural capacity of the beam (which is calculated as described in section 4.2.4 so as to allow for triaxial effects in the compressive zone). On the other hand, for the types of behaviour II and III, the assessment of load-carrying capacity may be based on the analytical description of the corresponding portions of the diagram of Fig. 3.9, as discussed below. (The assessment of the load-carrying capacity for type IV behaviour is discussed later.)

An analytical description of the portion of the diagram (a schematic representation of which is given in Fig. 4.2) corresponding to type II behaviour may be given by expression (4.1) presented below, which is a slightly modified (more conservative) version[1.8] of an empirically derived formula[2.22, 2.23] for a beam's shear capacity, the latter already being adopted by the British code of practice for the fire-resistant design of concrete structures:[4.1]

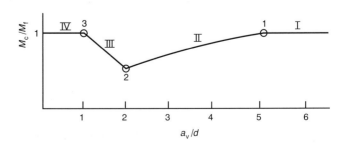

Fig. 4.2. Schematic representation of the failure criterion for the physical model in Fig. 4.1

$$M_c = 0{\cdot}875\, s\, d [0{\cdot}342 b_1 + 0{\cdot}3(M_f/d^2)\sqrt{(z/s)}] \sqrt[4]{16{\cdot}66/(\rho_w f_y)} \tag{4.1}$$

where s is the distance of the cross-section where M_c is calculated measured from the support closest to it:

$$s = \begin{cases} a_v & \text{for the case of two-point loading (i.e. the shear span); (single-point loading can be viewed as a special case of two coincident point loads)} \\ 2d & \text{for the case of uniformly-distributed loading (including any case where more than two point loads are uniformly distributed along the beam span)} \end{cases}$$

M_f is the flexural capacity of the section where M_c is calculated

d is the distance of the centroid of the tension reinforcement from the extreme compressive fibre (i.e. the effective depth)

z is the distance between the centroids of the compressive zone (assuming the rectangular stress block shown in Fig. 4.13 and allowing for any compression steel) and the tension reinforcement; clearly, z is the lever arm and it is always computed for the case of pure flexural action (even in the presence of thrust)

ρ_w = $A_s/(b_w d)$

b_w is the minimum breadth of the web (see Fig. 4.3)

f_y is the characteristic strength of the tension reinforcement

b_1 = min $(b_w + 2b_s,\ b_w + 2d_s)$ (see Fig. 4.3); (it should be noted that for a rectangular cross-section of width b, $b_1 = b$.)

Expression (4.1), in which the parameters entered must have the dimensions of N and/or mm, can also be used for the analytical description of the portion of the diagram in Fig. 4.2 representing type III behaviour. The position of any point within the latter regime may be assessed by interpolating linearly between the two end points of the portion corresponding to type III behaviour (points 2 and 3 in Fig. 4.2). The assessment of the position of these two points is relatively simple, since the co-ordinates of point 2 are $a_v/d = 2$ and M_c which is obtained from

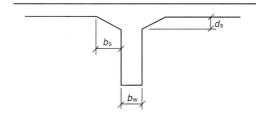

Fig. 4.3. Parameters b_s, b_w, and d_s used in equation (4.1)

expression (4.1) for $a_v/d = 2$, while the co-ordinates of point 3 are $a_v/d = 1$ and $M_c = M_f$ (see Fig. 4.2).

The effect of external axial load (H) on M_c may be allowed for by replacing parameter s in expression (4.1) with $s_0 = s(x_{M,H}/x_M)$, where $x_{M,H}$ and x_M represent the depth of the neutral axis of the beam (i.e. the depth of the horizontal element of the 'frame' of the proposed model) corresponding to the combined action of bending moment (M) and axial force (H) and to the action of the bending moment alone, respectively. As uncracked concrete in the compressive zone is the sole contributor to shear capacity, the introduction of s_0 in expression (4.1) describes in a sensible and simple manner the effect of changes in the size of the compressive zone on the load-carrying capacity of the beam. (For example, the existence of an H increases or reduces the size of the compressive zone depending on whether H is compressive or tensile respectively, thus leading to an increase or a reduction of the beam's load-carrying capacity.) It should be noted, however, that the practical significance of the above modification is small, as in most practical situations where a beam is also subjected to axial loading, expression (4.1), in both its initial (with s) and the modified (with s_0) form, is likely to lead to values of M_c larger than M_f, thus indicating type I behaviour. (For example, in columns, a_v/d is usually very large, so that M_c tends to be larger than M_f; the presence of an axial force makes M_c larger still while M_f may become smaller.)

As described in section 3.3, although the load-carrying capacity of beams with type IV behaviour corresponds to flexural capacity, loss of load-carrying capacity may be attributable to either failure of uncracked concrete in compression within the middle portion of the beam, i.e. failure of the horizontal element of the 'frame' of the proposed model, or failure in compression of the uncracked end portion of the beam, i.e. failure of the inclined leg of the 'frame' of the proposed model. In the former case, failure may be said to be ductile (resembling flexural failure), while in the latter case, it is brittle (resembling uniaxial compression failure). As brittle failure is undesirable, the prediction of only the load-carrying capacity is not adequate for design purposes; the prediction must also include the mode of failure. A simple design method that can also be used for predicting both load-carrying capacity and mode of failure for the case of two-point loading may be as follows (see also Fig. 4.4).[4.2]

> (a) Assuming the beam depth d and width b are given, the depth of the horizontal element of the 'frame' is assessed by satisfying the moment-equilibrium condition of the free body of Fig. 4.4 with respect to the intersection of the directions of the reaction and the longitudinal reinforcement. (If this condition cannot be satisfied (i.e. x is negative) with the given values of d and b, d and b are adjusted accordingly.) For purposes of determining the location of

DESIGN METHODOLOGY

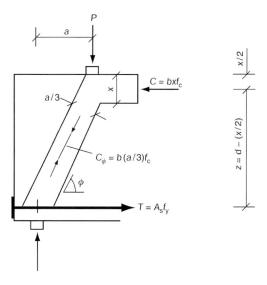

Fig. 4.4. Assessment of load-carrying capacity of beam with type IV behaviour

(a) Moment equilibrium $Cz = Pa$ yields x

(b) Horizontal force equilibrium $T = C$ yields A_s

(c) Check whether $a/3$ satisfies the condition $C_\phi \sin\phi > P$ which prevents brittle failure (where $\tan\phi = z/a$)
If not, adjust (b) and repeat

the joint between horizontal and inclined members of the 'frame' for type IV behaviour, the rules for type III behaviour can be taken to apply for both, two-point and uniformly-distributed loading, thus extending these rules to the range $a_v/d \leq 1$ and (equivalently) $L/d \leq 4$ respectively.

(b) Considering that the tension reinforcement yields before the load-carrying capacity of the horizontal element is attained (which is an input design aim), the amount of longitudinal reinforcement required is assessed by the equilibrium condition of the horizontal actions.

(c) Brittle failure is prevented when the vertical component of the compressive force carried by the inclined leg of the 'frame' is greater than, or equal to, the external load. If this condition is not satisfied, b is adjusted and the procedure is repeated. (It should be noted that the depth of the inclined leg is taken as equal to $a/3$ (where a is the shear span) as recommended in reference 4.1.)

Finally, it should be pointed out that the 'frame' models for behaviour types II, III and IV possess a horizontal member irrespective of load type (two-point or uniformly-distributed loading). The exceptions occur for behaviour types III and IV in the limiting case of single-point loading, in which instance the 'horizontal' member shrinks to the joint of the two inclined legs.

4.2.3. Validity of failure criteria

The investigation of the validity of the failure criteria proposed is based essentially on a comparative study of predicted behaviour and experimental information on the load-carrying capacity and mode of failure of simply-supported reinforced concrete beams with a wide range of geometric characteristics and loading conditions. Figures 4.5 to 4.12 depict the results of such a comparative study using a large amount of experimental information; Figs 4.5 to 4.7 also include typical predictions stemming from the methods currently used for the design of concrete structures.

Figure 4.5 refers to two types of beam with a rectangular cross-section and similar geometric characteristics, with values of the similitude ratio between 1 and 4. The beams are subjected to two-point loading with a normalised value of the shear span (i.e. shear span-to-depth ratio (a_v/d)) equal to 3, corresponding to type II behaviour. From the figure, it becomes apparent that both the proposed failure criteria and the current design methods yield predictions which correlate closely with the experimentally established behaviour, the latter stemming from reference 3.2.

Figure 4.6 also refers to two types of beam, with the same cross-section in respect of both the geometry (i.e. rectangular shape, dimensions, reinforcement, etc.) and the quality of concrete and steel.[3.2] The beams in Fig. 4.6(a) have a span of 6·0 m and they are subjected to two-point loading with varying shear span. Depending on the length of the shear span the beams may exhibit either type II (for values of the shear span larger than 0·54 m) or type III behaviour (for values of the shear span smaller than 0·54 m) (see Fig. 4.2). On the other hand, the beams in Fig. 4.6(b) have a span varying between 1·5 m and 6·0 m, subjected to a uniformly-distributed load, and, except for those with a span smaller than 2·16 m which are characterised by type III behaviour, exhibit type II behaviour (see Figs 4.1 and 4.2). From the figures, it becomes apparent that the proposed failure criteria yield predictions which are significantly closer to the experimentally established behaviour (as reported in reference 3.2) than the predictions of current design methods. The superiority of the proposed failure criteria become more evident with a reduction of both the shear span, for the case of two-point loading, and the span, for the case of uniformly-distributed loading.

This superiority of the proposed failure criteria (i.e. for type II and type III beams) becomes really striking for the case of beams with a T-shaped section.[2.21] Figure 4.7(a) indicates that, in contrast with the predictions of the proposed failure criterion which correlate closely with the experimental behaviour, the predictions of current design methods underestimate considerably the experimental values. For the case of beams with a span of 2·6 m, the value of the load-carrying capacity predicted by the code is approximately 50% of that established by experiment for the

DESIGN METHODOLOGY

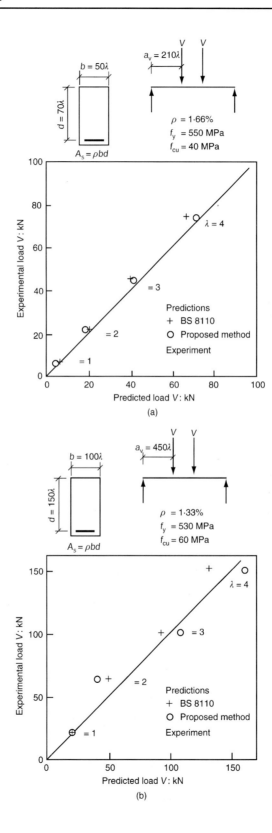

Fig. 4.5. Load-carrying capacity of beams with the same geometric characteristics and different similitude ratios λ between 1 and $4^{3.2}$

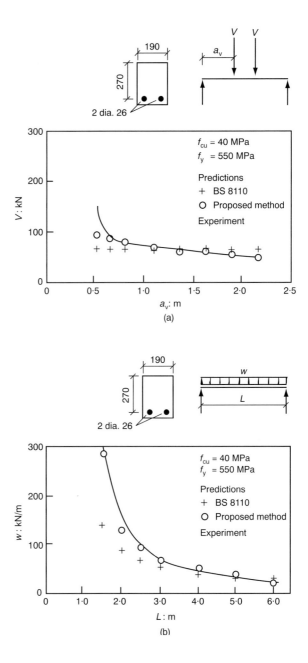

Fig. 4.6. Load-carrying capacity of beams with type II behaviour under (a) two-point loading and (b) uniformly-distributed loading[3.2]

case of two-point loading, and as low as 20% for the case of multiple-point loading (which is essentially equivalent to uniformly-distributed loading).

Here, one must emphasise the significance placed by the proposed failure criterion for type II behaviour on the shape of the compressive zone which is completely ignored by current design methods.[2.23] The importance of the shape of the compressive zone is related to the development of lateral tensile

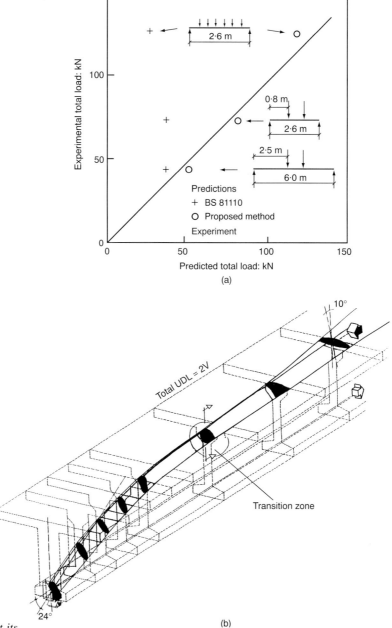

Fig. 4.7. T-beam at its ultimate limit state: (a) load-carrying capacity[2.21] and (b) schematic representation of the beam portion enclosing the compressive-stress trajectories[2.23]

stresses in the region where the flange connects with the web at the location where the internal compressive stress resultant changes direction, which, for the case of type II behaviour, is located at a distance equal to $2d$ from the support. As indicated in Fig. 4.7(b),[2.23] the shape of the cross-section essentially dictates the shape of the distribution of the internal compressive stresses.

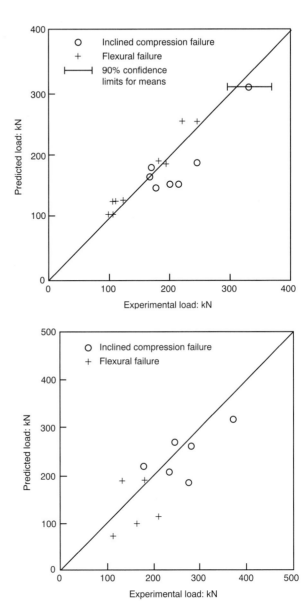

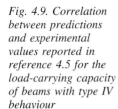

Fig. 4.8. Correlation between predictions and experimental values reported in references 4.3 and 4.4 for the load-carrying capacity of beams with type IV behaviour

Fig. 4.9. Correlation between predictions and experimental values reported in reference 4.5 for the load-carrying capacity of beams with type IV behaviour

In fact, the distribution of the compressive stresses within the web in the region close to the support has a significantly larger spread in the vertical rather than in the lateral direction, whereas the larger spread changes direction within the flange and becomes lateral in the region of the beam beyond the cross-section located at a distance equal to $2d$ from the support. Such a change in the direction of the larger spread of the compressive stresses indicates that, in the region of the cross-section located at a distance of approximately $2d$ from the support, the compressive-stress trajectories change direction not only in the vertical (as already

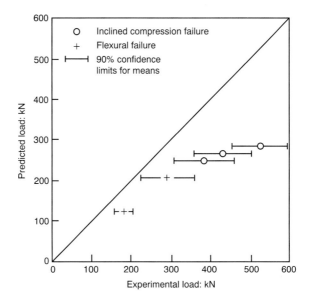

Fig. 4.10. Correlation between predictions and experimental values reported in reference 4.6 for the load-carrying capacity of beams with type IV behaviour

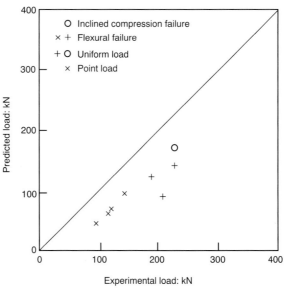

Fig. 4.11. Correlation between predictions and experimental values reported in reference 4.5 for the load-carrying capacity of beams with type IV behaviour

discussed in section 3.2.5) but also in the lateral direction. Such changes in the direction of the compressive-stress trajectories will cause the development of transverse tensile stresses in both the vertical and lateral directions.

The abrupt change in the lateral direction of the stress trajectories may be smoothened considerably if a transition is formed between the flange and the web, as indicated in Fig. 4.7(b), thus reducing the size of the transverse tensile stresses in the horizontal direction. The proposed failure criterion for type II behaviour allows for the dependence of the internal stress field on the shape

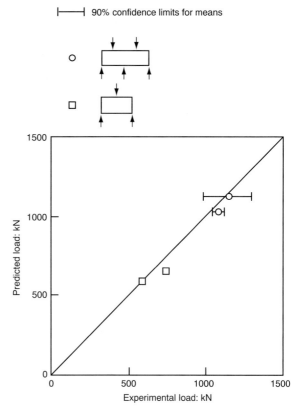

Fig. 4.12. Correlation between predictions and experimental values reported in reference 4.7 for the load-carrying capacity of beams with type IV behaviour

of the beam's cross-section, especially the beneficial effect of tapering, and hence leads to a significant improvement of the predictions of the load-carrying capacity of beams with T-sections.

Figures 4.8 to 4.12 refer to deep beams, i.e. beams characterised by type IV behaviour. The figures provide an indication of the relationship between the predicted and experimental[4.3–4.8] values of the load-carrying capacity, together with predictions of the mode of failure. From the figures, it becomes apparent that the predictions of the proposed failure criterion are satisfactory for all the cases investigated, in spite of the simplifications incorporated in the proposed criterion summarised in Fig. 4.4.

4.2.4. Assessment of longitudinal reinforcement

For a cross-section with given geometry (i.e. given shape and dimensions) subjected to a bending moment M_d, the longitudinal reinforcement may be assessed such that the flexural capacity M_f is at least equal to the acting bending moment M_d. The method of assessment (which is slightly different from that incorporated in current codes with regard mainly to the intensity of the compressive stress block, now taken equal to f_{cyl} as a more realistic average value to allow for the effect of the triaxial stress conditions, rather than a value of approximately $0.85 f_{cyl}$ which is specified by most current codes) is described in Fig. 4.13.

DESIGN METHODOLOGY

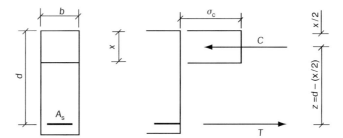

Fig. 4.13. Assessment of flexural capacity for the physical model of an RC beam

$M_f = T * z = C * z$
where $C = b * x * \sigma_c$
        $T = A_s * f_y/\gamma_s$
        $\sigma_c = f_{cyl}/\gamma_c$
        f_{cyl} = cylinder concrete strength
        f_y = characteristic strength of tension steel
        γ_c, γ_s = partial safety factors for concrete and steel

The figure provides a simplified description of the internal actions developing at a given cross-section just before it fails in flexure. From the figure, it appears that the compressive zone of the beam (horizontal element of 'frame' of proposed physical model) may be considered to be subjected to a uniform stress which yields a stress resultant equivalent to that corresponding to the actual stress distribution that develops due to the bending of the beam. The intensity of the uniform stress block is taken equal to the strength of concrete in uniaxial compression (f_{cyl}), as established from tests on cylinders with a value of the height-to-diameter ratio between 2 and 2·5.[1.10] Such a value of the stress intensity, although larger by approximately 15% than that used by current design methods, is still considered to be conservative, since the experimental information presented in Chapter 2 indicated that, owing to the triaxiality of the stress conditions, the average stress that develops in the compressive zone of a beam at its ultimate limit state in flexure is significantly larger than the uniaxial compressive strength of concrete. Thus, the larger 'equivalent' uniaxial strength proposed constitutes an attempt to compensate partly, and in a simple manner, for the ignoring of unavoidable triaxial-stress conditions which only a complex three-dimensional finite-element analysis, such as the one described in reference 1.10, can establish.

The compressive stress resultant, acting on the horizontal element of the 'frame' of the proposed physical model, is $C = bx(f_{cyl}/\gamma_c)$, while the tensile force sustained by the reinforcement is $T = A_s(f_y/\gamma_s)$, where x is the depth of the horizontal element (compressive zone of beam), d is the distance of the centroid of the longitudinal steel bars from the top face of the member, A_s is the area of the cross-section of the bars, f_y is the characteristic strength of the reinforcement, and γ_c, γ_s are the partial safety factors for concrete and steel respectively. The equivalence between internal and external actions yields the following equations:

$$C[=bx(f_{cyl}/\gamma_c)] = T[=A_s(f_y/\gamma_s)] \quad (4.2)$$

$$M_f = Cz \quad (4.3)$$

where $z = d-x/2$ is the distance between the points of application of C and T.

By replacing in the latter of the above equations (i.e. 4.3) $C = bx(f_{cyl}/\gamma_c)$ (i.e. 4.2), $M_f = M_d$, and $z = d-x/2$, a quadratic equation in x is obtained which yields

$$x = d\{1 \pm \sqrt{1 - [2\gamma_c M_d/(bf_{cyl} d^2)]}\} \quad (4.4)$$

As $x<d$, only the minus sign has physical significance. Moreover, if $2\gamma_c M_d/(bf_{cyl} d^2) > 1$, b and d must be reassessed so that $2\gamma_c M_d/(bf_{cyl} d^2)$ becomes smaller than 1.

Having established the value of x, equation (4.2) yields the value of T, which then enables the calculation of the amount of reinforcement to be obtained from the expression $A_s = T/(f_y/\gamma_s)$.

4.2.5. Assessment of transverse reinforcement

For behaviour of types II and III, transverse reinforcement may be used in order to prevent failure from occurring before flexural capacity is exhausted. In all other cases, a nominal amount of transverse reinforcement is deemed sufficient for sustaining tensile stresses of the order of 1 MPa (which, in most cases, is somewhat more conservative than the value specified in current codes).

As discussed in section 3.3, the most likely mode of failure of beams characterised by type II behaviour involves the near-horizontal splitting of the compressive zone of the beam in the region where the path of the compressive stress resultant changes direction (i.e. the separation of the 'frame' from the remainder of the 'comb-like' model in the region of the junction of its horizontal and inclined elements). Such a splitting is essentially caused by the vertical component of the inclined compressive stress resultant that develops within the beam end (i.e. within the inclined leg of the 'frame' of the proposed model), this vertical component being equivalent to the shear force acting in the region of the change in the force-path direction. (As explained in section 3.2.5, this can also be understood by viewing the 'kink' in the stress path as giving rise to an orthogonal force bisecting the angle between the horizontal portion and the inclined leg of the frame: that such a force is tensile is evident from the fact that it tends to separate the compressive zone of the beam from the cracked region below it.)

The maximum value of the shear force that can be sustained by concrete in the above region easily results from expression (4.1). For example, the case of two point-loading (which also

DESIGN METHODOLOGY

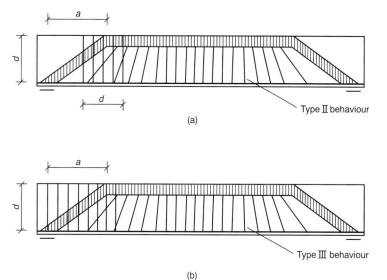

Fig. 4.14. Schematic representation of transverse reinforcement for RC beams exhibiting types of behaviour (a) II and (b) III

Note: Only 'calculated' transverse reinforcement is indicated. Throughout the remainder of the beam, nominal transverse reinforcement is provided in compliance with current design practice.

includes the case of single-point loading, as the latter may be viewed as a two-point loading for which both loads are applied at the same point), this maximum value is $V_c = M_c/s$, while for the case of uniformly-distributed loading, the load that corresponds to bending moment M_c at the cross-section $s = 2d$ is $q_c = M_c/[d(L-2d)]$ and, therefore, the value of the shear force is $V_c = M_c(L-4d)/[2d(L-2d)]$, where M_c, in both cases is calculated from equation (4.1).

Failure (before flexural capacity is exhausted) will occur when $M_c < M_f$ or $V_c < V_f$ (where V_f is the value of the shear force corresponding to the flexural capacity M_f). This type of failure may be prevented by placing transverse reinforcement sufficient to sustain the portion of V_f, in excess of V_c that can be sustained by concrete alone. If the characteristic strength of the transverse reinforcement is f_{yv}, the amount of reinforcement required is: $A_{sv} = (V_f - V_c)/(f_{yv}/\gamma_s)$. Such reinforcement is placed in the region of the joint or junction of the horizontal and inclined elements of the 'frame' of the proposed physical model, uniformly distributed within a length equal to the beam depth, and symmetrically situated about the joint as indicated in Fig. 4.14(a). (Fig. 4.14(b) refers to type III behaviour, to be discussed subsequently.) In the remainder of the beam it is deemed sufficient to place nominal reinforcement capable of sustaining tensile stresses of the order of 1·0 MPa. Such nominal reinforcement should be placed throughout the beam span when $M_c > M_f$ or $V_c > V_f$, and, hence, there is no need to specify additional transverse reinforcement in the region of the joint.

Fig. 4.15. State of stress of the beam portion between cross-sections including consecutive cracks after the loss of bond between concrete and longitudinal reinforcement (Note: for non-rectangular stress blocks, the lever arms for C, x/2 and x_1/2, should be replaced by the relevant centroidal distances of the compressive force)

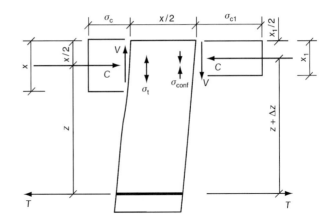

As discussed in section 3.2.5, for the case of a beam subjected up to two point loads, the causes of failure characterising type II behaviour may be associated with the loss of bond between concrete and the longitudinal reinforcement in the region of the cross-section that includes a point load (see item (d) in section 3.2.5). (Such bond failure is only associated with type II behaviour and, say, two-point loading, since uniformly-distributed loading results in small shear forces throughout much of the central span: for more than two point loads, an equivalent uniformly-distributed loading can be assumed but, if in doubt, a check for shear can always be made for a (small) finite number of point loads.) Figure 4.15 illustrates a portion of the beam between cross-sections which include two consecutive cracks, together with the internal actions developing on these sections after the loss of bond.

The manner in which the loss of bond causes the change in the internal actions and leads to the development of transverse actions within the compressive zone of the element, as indicated by σ_t in Fig. 4.15, is described fully in section 3.2.5 and Fig. 3.8 of the preceding chapter. An assessment of the value of the transverse tensile stresses, in excess of those that can be sustained by concrete alone within the compressive zone of the element, may be made by following the steps described below (refer also to Fig. 4.15).

(a) Calculate the increase Δz of the lever arm z of the internal actions $C = T$ at the right-hand side of the element. From the equilibrium condition $Va = T\Delta z$ results $\Delta z = Va/T$, where $V = V_d - V_c$ (with V_d being the design shear force and $V_c = M_c/s$ the shear force sustained by concrete alone), and $a = x/2$ is a value derived on the basis of the experimental data of reference 3.2 (with x, as indicated in Fig. 4.1, being the depth of the compressive zone).

(b) Having established Δz, the depth (x_1) of the compressive zone at the right-hand side of the element is easily found to be $x_1 = 2(d - z - \Delta z)$.

(c) The mean compressive stress at the right-hand side of the element is given by $\sigma_{c1} = C/(bx_1)$. (This assumes a rectangular stress block or a rectanglar approximation to more complex stress blocks (as in T-beams in which the compressive zone also includes trapezoidal or stepped shapes); strictly, of course, (bx_1) should be replaced by the actual area of the new compressive zone.)

(d) For $\sigma_{c1} > f_{cyl}$ (where f_{cyl} is the maximum value of the average compressive stress that can be developed in the compressive zone), σ_{c1} can only be developed in the presence of a confining transverse stress σ_{conf} which may be estimated on the basis of the triaxial strength envelope of Fig. 2.11 as $\sigma_{conf} = (\sigma_{c1} - f_{cyl})/5.1^{.8}$

(e) The confining pressure σ_{conf}, set up on the right-hand side of the element as a result of local dilation there, is provided by the adjacent concrete to its left. As explained in section 2.3.2.1, the dilating region induces tensile stresses in the restraining region. The mean value of this transverse tensile stress σ_t which develops at the upper left-hand side of the element is $\sigma_t = \sigma_{conf}$.

After the value of the average tensile stress has been established, beyond that that can be sustained in the transverse direction by concrete alone, the amount of reinforcement required to sustain it within a distance d (within the shear span) from the point of application of the point load is given by $A_{sv} = \sigma_t b d/(f_{yv}/\gamma_s)$. However, it is suggested, as 'good practice', that this distance d (over which the stirrups should, in theory, suffice) be extended by placing this flange reinforcement throughout the horizontal member of the 'frame' within the shear span while also extending it within the flexure span to a distance d from the cross-section at which the point load acts. Clearly, the stirrups in the compression zone of the beam must be distributed across the beam's width so that the usual vertical stirrups running through the beam's depth are ineffective (especially in the case of flanges of T-sections). Moreover, if such distributed reinforcement is confined to vertical legs, horizontal splitting would be prevented while the vertical splitting would only be delayed but not prevented, with the result that, although significant ductility might be achieved, the full flexural capacity will not be reached. All this suggests the need to reinforce the compressive zone/flange of a beam with hoop stirrups so that the tensile stresses resulting from triaxial dilation perpendicular to the direction of principal compressive stress are fully taken up in both horizontal and vertical directions, thus enabling the member to attain its full flexural capacity. (Note that no confining role is intended by using hoop stirrups, as the spacing required for effective confinement would be much

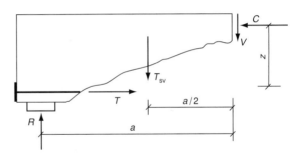

Fig. 4.16. Assessment of transverse reinforcement of beams with type III behaviour

closer than that presently needed for simply catering for tensile stresses.)

As discussed in section 3.3, failure characterising type III behaviour is attributable to the reduced strength of the compressive zone of the uncracked portion of the beam (i.e. the horizontal element of the 'frame' of the proposed model) adjacent to the region of the change in the path of the compressive stress resultant (i.e. adjacent to the region of the joint of the horizontal and inclined members of the 'frame'). This strength reduction, which is due to the deep penetration of the inclined crack closest to the support into the compressive zone (see Fig. 3.1), results in a reduction of the beam's flexural capacity.

The beam's load-carrying capacity may be increased to the value corresponding to the flexural capacity by uniformly distributing transverse reinforcement within the whole length of the horizontal projection of the inclined leg of the 'frame' (see Fig. 4.14(b)) (i.e. of the horizontal projection of the inclined crack closest to the support (see Fig. 3.1)). Figure 4.16 depicts the portion of the beam in Fig. 3.1 enclosed by its left-hand end-face, the deepest inclined crack closest to the support, and the cross-section of the horizontal element with the reduced strength. If it is assumed that the transverse reinforcement is at yield, then the total force that can be sustained by such reinforcement is $T_{sv} = A_{sv}(f_{yv}/\gamma_s)$ which acts in the middle of the length a of the portion considered. For the equilibrium of this portion, $Ra - T_{sv}(a/2) - Tz = 0$, where $M_c = Tz$ and $M_f = Ra$, while T_{sv} is the force which must be sustained by the transverse reinforcement for the flexural capacity of the cross-section at a distance a from the support to increase from M_c to M_f. As a result, $T_{sv} = 2(M_f - M_c)/a$ (and, for two-point loading, $a = a_v$), and thus the total amount of reinforcement required to sustain this force is $A_{sv} = T_{sv}/(f_{yv}/\gamma_s)$. In other words, while the codes recommend stirrups to complement the shear capacity in the shear span (leading, in the case of two-point loading, to $V_s = V_f - V_c = (M_f - M_c)/a$ (where, again, $a = a_v$)), the present approach concentrates on the deleterious effect the shear crack has on flexural capacity (making $M_c < M_f$) and, to offset this, proposes the introduction of stirrups to compensate for the reduced cross-section moment; as shown above, for the case of two-point loading, this requires twice the amount of transverse reinforcement recommended by codes. For the case of uniformly-distributed loading, it is easy to show that, while the code requires $T_{sv} = 2(M_f - 4/3\, M_c)/L$, the present approach demands almost four times this force (and hence stirrup reinforcement), the actual value being $T_{sv} = 8(3/4\, M_f - M_c)/L$. (The factors 4/3 and 3/4 appear because the flexural capacity refers to the midspan of the beam and not the 'critical' cross-section at a distance $a = L/4$ from the support.)

4.2.6. Design procedure

The design of a reinforced concrete beam involves, on the one hand, the selection of materials (concrete and steel) of a suitable quality, and, on the other hand, the determination of the geometric characteristics (i.e. shape and dimensions) of the member, inclusive of the amount and location of reinforcement, required for the beam to have a given load-carrying capacity and ductility. However, the selection of the quality of the materials is independent of the proposed methodology, since the underlying theory is valid for the whole range of material qualities available to date for practical applications. Also, out of the geometric characteristics, the span of the beam may be considered as known, since it results directly from the overall structural configuration adopted. Hence, the design of a simply-supported reinforced concrete beam involves essentially the determination of the cross-sectional characteristics (i.e. shape and dimensions) of the member together with the amount and location of the reinforcement.

Bearing in mind the above, the information presented in sections 4.2.1 to 4.2.5 may be incorporated into the following design procedure which comprises six steps:

(a) *Preliminary assessment of geometric characteristics.* This may be carried out by following current design practice as described in reference 2.4. For example, with the exception of deep beams (i.e. beams characterised by type IV behaviour), the cross-sectional depth (d) is taken approximately equal to $L/12$, where L is the beam span, while the web width (b_w) of the cross-section is given a value between $d/3$ and $2d/3$. (For a rectangular cross-section, $b = b_w$.) For the case of deep beams, the beam depth is such as to satisfy the condition $L/2d < 1$, which defines deep-beam behaviour, while the width may be taken initially to be equal to $L/24$.

(b) *Calculation of design bending moment and shear force.* With the applied load and the beam span known, the bending-moment and shear-force diagrams may be easily constructed. The design bending moment and shear force at 'critical' cross-sections are obtained from these diagrams.

(c) *Assessment of longitudinal (flexural) reinforcement.* If it is assumed that the flexural capacity (M_f) is at least equal to the design bending moment (M_d), the amount (A_s) of longitudinal reinforcement required is calculated as described in section 4.2.4. (It should be noted from Fig. 4.13 that the calculation of A_s is preceded by the calculation of the depth x of the beam's compressive zone (i.e. the depth of the horizontal element of the model 'frame'). (As remarked earlier, if the resulting value of x is not a real number, then the geometric characteristics must be reassessed such that the value of x becomes real.)

(d) *Construction of physical model.* As indicated in Fig. 4.1, the shape of the model is essentially given; only the position of the joint of the horizontal and inclined members of the 'frame' must be determined. Figure 4.1 defines the above position for the cases of two-point loading (which also describes the case of single-point loading if the two point loads have a common point of application) and uniformly-distributed loading. For a number of point loads larger than two, point-loading may be considered equivalent to uniformly-distributed loading with the same total load.

(e) *Determination of type of beam behaviour.* From Fig. 3.9 in the preceding chapter, the type of behaviour may determined by the value of a_v/d or L/d depending on the type of applied load.

(f) *Calculation of transverse reinforcement.* Transverse reinforcement is required only for behaviour of types II and III. For all other cases, it is sufficient to provide nominal reinforcement capable of sustaining tensile stresses of the order of 1·0 MPa. For the case of type II behaviour, the method of calculation of transverse reinforcement (required for all types of loading at the location of the joint of the horizontal and inclined elements of the model's frame) is described in section 4.2.5. This section also describes the method of calculation of the additional transverse reinforcement that may be required to prevent horizontal splitting of the compressive zone for the case of point-loading. Throughout the remainder of the beam, nominal reinforcement is provided as for the cases I and IV. For type III behaviour, the calculation of transverse reinforcement is carried out as described at the end of section 4.2.5.

4.2.7. Design examples

In what follows, the first example shows an instance of design, while in the other two examples, checks are made on existing arrangements.

DESIGN METHODOLOGY

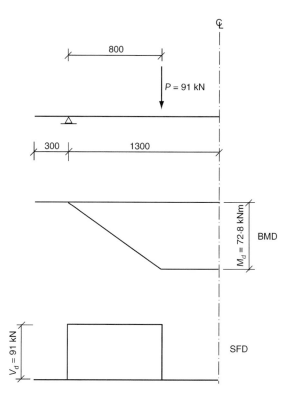

Fig. 4.17. Longitudinal dimensions, position and magnitude of design point load(s), and corresponding bending-moment and shear-force diagrams of a beam with the cross-section shown in Fig. 4.18[1.8]

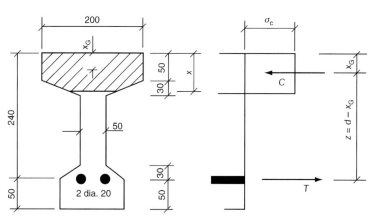

Fig. 4.18. Cross-sectional geometric characteristics of the beam under two-point loading depicted in Fig. 4.17[1.8]

(a) Beam of type II behaviour

Figures 4.17 and 4.18 illustrate the geometric characteristics of a 'laboratory' beam under two-point loading, symmetrical with respect to the mid cross-section. The behaviour of the beam was established experimentally,[1.8] the aim being that of verifying the validity of the proposed methodology.

The beam was constructed by using concrete with a characteristic strength in uniaxial compression $f_c = 26$ MPa, longitudinal steel bars with yield stress $f_y = 560$ MPa, and stirrups with a yield stress $f_{yv} = 450$ MPa. All safety factors were taken equal to 1. In the following, the design procedure described in the preceding section is applied in order to calculate the amount of reinforcement required for the beam to have a load-carrying capacity equal to $2P = 2 \times 91 = 182$ kN.

Design actions
Figure 4.17 depicts the diagrams of the internal actions corresponding to a load-carrying capacity of 182 kN. The values of the design bending moment (M_d) and shear force (V_d) obtained from the diagrams are 72·8 kNm and 91 kN, respectively.

Longitudinal reinforcement
A preliminary assessment of the amount of longitudinal reinforcement required may be made by assuming that the lever arm of the longitudinal internal actions is $z = 0.85 \times d = 204$ mm (with d assumed at $d = 240$ mm). As the flexural capacity is given by $M_f = Tz = A_s f_y z = M_d$ (see Fig. 4.18), it follows that $A_s = M_d/(f_y z) = 72\,800\,000/(560 \times 204) = 637$ mm². Such a value of A_s is equivalent to 2 dia. 20 ($= 628$ mm²) which are placed so that their geometric centre is located at a distance equal to 50 mm from the extreme bottom fibre of the cross-section (as implied in the d value assumed above).

Verification of adequacy of reinforcement. For 2 dia. 20, the total tensile force sustained by the two bars is $T = 628 \times 560 = 351\,680$ N. Since $T = C = A_c \sigma_c$, where A_c is the cross-sectional area of the compressive zone (see Fig. 4.18), $A_c = C/\sigma_c = 351\,680/26 = 13\,526$ mm². For the given shape of the cross-section, the depth of the compressive zone is $x = 76$ mm, while its geometric centre is located at a distance equal to $x_G = 34$ mm. Hence, $z = 240 - 34 = 206$ mm and $M_f = Tz = 351\,680 \times 206 = 72\,400\,000$ (Nmm) $= 72.4$ (kNm), the latter value being very close to the design value $M_d = 72.8$ kNm.

Physical model
For the case of two-point loading, the position of the joint of the horizontal and inclined elements of the 'frame' of the model depends on the value of a_v/d. Since $a_v/d = 800/240 = 3.33 > 2$, the beam is characterised by type II behaviour and, hence, the distance of the joint from the support is equal to $2d = 2 \times 240 = 480$ mm (see Fig. 4.19).

Transverse reinforcement
For type II behaviour, expression (4.1) gives $M_c = 33.62$ kNm which results in $V_c = M_c/s = M_c/a_v = 42$ kN. The

DESIGN METHODOLOGY

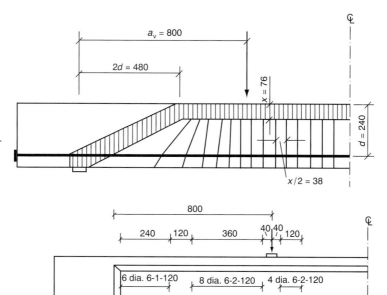

Fig. 4.19. Details of the physical model for the beam with the characteristics shown in Figs 4.17 and 4.18

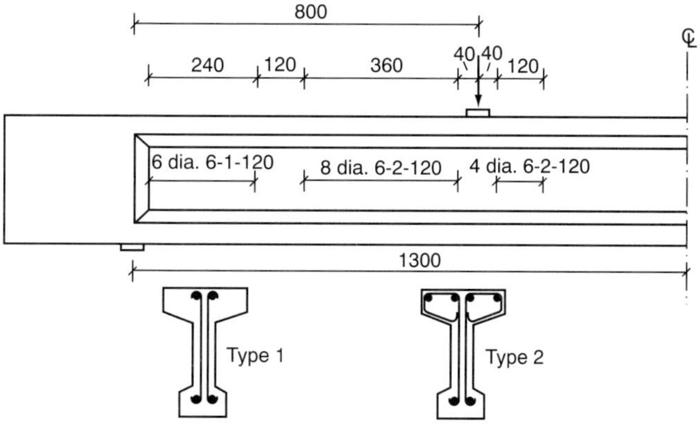

Fig. 4.20. Transverse-reinforcement details for the beam in Figs 4.17 and 4.18

transverse reinforcement required in the region of the 'joint' in order to sustain the shear force $V_s = V_d - V_c = 49$ kN is $A_{sv} = V_s/f_{yv} = 49\,000/450 = 109$ mm.^{2} (Note that, although V_c is computed at $s = a_v$, the same shear force acts at the joint of the 'frame' in the case of two-point loading.) Such reinforcement is equivalent to two two-legged stirrups dia. 6 (=113·04 mm^2) which, as described in section 4.2.5, is placed within a distance of d (= 240 mm) symmetrically located with respect to the 'joint,' as indicated in Fig. 4.20.

Transverse reinforcement may also be required within the compressive zone of the shear span in the region of the point load in order to sustain the tensile stresses that develop as a result of failure of the bond between concrete and the longitudinal reinforcement (see section 3.2.5). Following the procedure described in section 4.2.5 and by reference to Fig. 4.15, the reinforcement required is determined by assessing the following parameters:

$$\Delta z = Vx/(2T) = 49\,000 \times 76/(2 \times 351\,680) \approx 5 \text{ mm}$$
$$x_1 = 2(d - z - \Delta z) = 2 \times (240 - 206 - 5) = 58 \text{ mm}$$
$$\sigma_{c1} \approx C/(bx_1) = 351\,680/(200 \times 58) = 30\cdot3 \text{ MPa}$$

(note that, here, the use of (bx_1) constitutes an approximation to the slightly non-rectangular stress block implicit in $x_1 = 55$ mm)

$\sigma_{conf} = (\sigma_{c1} - f_{cyl})/5 = (30\cdot3 - 26\cdot0)/5 = 0\cdot86$ MPa
$\sigma_t = -\sigma_{conf}$ (compression $= +$ ve)
$T_{sv} = \sigma_t bd = -0\cdot86 \times 200 \times 240 = -41\,280$ kN
$A_{sv} = 41\,280/450 = 92$ mm^2

Two two-legged stirrups dia. 6 (2 dia. $6 = 113\cdot04$ mm^2) are placed within a distance of 240 mm ($= d$) from the point load within the compressive zone of the shear span of the beam, as indicated in Fig. 4.20. The figure shows that, in addition, this stirrup arrangement is then extended a further distance so as to cover the whole length of the horizontal member of the 'frame' within the shear span, as suggested in section 4.2.5: there, it was also proposed that an extension into the flexure span of d be made, whereas Fig. 4.20 shows a somewhat closer spacing near the point load (80 mm instead of 120 mm) but a slightly shorter extension (200 mm instead of 240 mm) because this particular beam was designed at a time when the present design methodology was still being developed.[1.8] Within the portion of the beam where the proposed model does not specify transverse reinforcement, a nominal number of stirrups is uniformly distributed sufficient to sustain tensile stresses of the order of 1 MPa, the vertical stirrup type adopted having the same size as the rest of the bars (see Fig. 4.20). From the experimental results provided in reference 1.8, it can be seen that the beam exhausted its flexural capacity, with the predicted value of the load-carrying capacity (181·8 kN) being only 10·4 kN smaller than the experimental value (192·2 kN); in fact, the beam failed by vertical splitting of the compressive zone along the longitudinal axis (as shown clearly in Fig. 10(b) of reference 1.8) as the horizontal legs of the flange reinforcement were not continuous throughout the flange width (see Fig. 4.20).

(b) **Beam of type III behaviour**
Figure 4.21 illustrates the geometric characteristics of a beam with type III behaviour (i.e. with the value of a_v/d between 1 and 2), under a single point load applied at mid-span. This beam is one of the members which were used to investigate the behaviour of beams with $a_v/d < 2$.[4.7] The beam was constructed with a concrete with $f_c = 42\cdot4$ MPa, six longitudinal bars with a total tensile force capacity $T_s = 546$ kN, and five two-legged stirrups

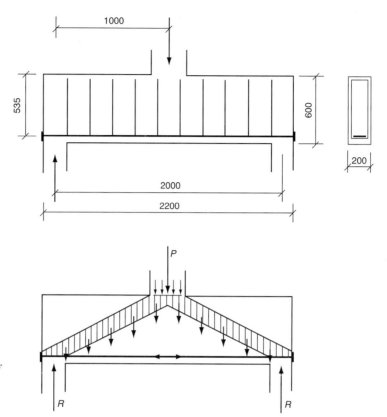

Fig. 4.21. Design details and position of point load for the beam with type III behaviour

with a total tensile force capacity $T_{sv} = 162$ kN. In the following, the proposed methodology is used to predict the mode of failure and the load-carrying capacity of the beam by using the physical model also shown in Fig. 4.21. Note that the single-point loading precludes the formation of a finite horizontal member of the 'frame' as explained at the end of section 4.2.2 (other instances of this will also be given later). An indication of the validity of the methodology is obtained from the comparison between predictions and experimental values.

Figure 4.22 depicts the internal actions which develop at the mid cross-section of the beam exhibiting a flexural mode of failure. Assuming safety factors equal to 1, the equilibrium condition $C = T$, where $C = bxf_c = 8480x$ and $T = 546\,000$ N, yields a depth of the compressive zone $x = 64$ mm, and thus the lever arm of the couple of the longitudinal internal actions $(C = T)$ is $z = d - x/2 = 535 - 32 = 503$ mm. The moment of this couple yields the beam's flexural capacity $M_f = Cz = Tz = 546 \times 0.503 = 274.64$ kNm.

Failure of the beam may also result from the reduction of the depth of the compressive zone owing to the extension of the inclined crack which is deeper than the flexural cracks. (The

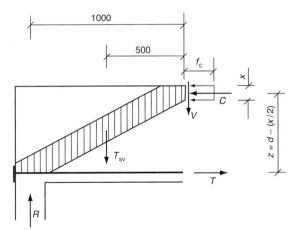

Fig. 4.22. Internal actions developing at the mid cross-section of the beam in Fig. 4.21 just before it fails in flexure

causes of this type of failure — occuring as a result of failure of the horizontal element of the 'frame', in the region of the joint of the horizontal and inclined elements of the 'frame' — were described in section 3.3 of the preceding chapter.) The reduction in depth causes a reduction in flexural capacity and the maximum bending moment that can be sustained by the beam can be assessed, as described in section 4.2.2, by linear interpolation between two values of M_c, $M_c = M_f$ for $a_v/d = 1$, and the value obtained from expression (4.1) for $s = 2d$ corresponding to $a_v/d = 2$. Thus for $a_v/d = 2$, expression (4.1) yields $M_c = 178·91$ kNm, while for $a_v/d = 1$, $M_c = M_f = 274·64$ kNm. As a result, for $a_v/d = 1·869$, the maximum moment that can be sustained by the combined resistance of the compressive zone and the longitudinal tension reinforcement is $M_c = 274·64 - 0·869 \times (274·64 - 178·91) = 191·45$ kNm.

As described in section 4.2.5 (see also Fig. 4.22), the total tensile force $T_{sv} = 162$ kN, sustained by (the uniformly distributed) transverse reinforcement, increases the above value of bending moment by $\Delta M_c = 162 \times 0·5 = 81$ kNm. Hence, for the case of the mode of failure considered, the maximum bending moment that can be sustained by the beam is $M_c = 191·45 + 81 = 272·45$ kNm.

Comparing the above value with the value of 274·64 kNm, which corresponds to flexural capacity, leads to the conclusion that the beam fails before its flexural capacity is exhausted. The load-carrying capacity of the beam is obtained by the equilibrium condition $Ra_v = M_c = 272·45$ kNm (with $a_v = 1·0$ m) or $R = 272·45$ kN; $P = 2R = 2 \times 272·45 = 545$ kN. The experimental data provided in reference 4.7 show that the beam did indeed fail in a brittle manner, with the predicted value of the load-carrying capacity being equal to approximately 77% of the experimentally established value. (The higher experimental value is due to the strain hardening of the longitudinal reinforcement which was ignored by the design method employed, as reference 4.7 does not provide any information on strain hardening. Had strain hardening been allowed, the design calculations would have yielded a higher value of M_f and, hence, a higher value of M_c since, as indicated in equation (4.1), M_c is a function of M_f. As a result, the predicted value would be closer to its experimental counterpart.)

(c) Beam of type IV behaviour

The deep beam (with $a_v/d \sim 1$) in Fig. 4.23 is one of the specimens of the experimental programme referred to in the preceding example$^{4.7}$ and it is used to verify the validity of the proposed design method for beams of type IV behaviour, described in Fig. 4.4. The beam was constructed with a concrete with $f_c = 26$ MPa and six longitudinal bars with total tensile force capacity of $6 \times 114 = 684$ kN. In the following, the proposed method is used to predict both the load-carrying capacity and the mode of failure of the beam.

As described in section 4.2.2, the load-carrying capacity of the beam depends on the compressive strength of the weakest element of the 'frame' of the model in Fig. 4.24. (It should be noted that, for the reasons given in the previous example, there is no finite horizontal member to the frame model.) From the figure, the depth x of the cross-section of the 'horizontal' element (in this instance, of course, the latter is made up entirely of the 'junction' in the 'frame') is obtained from the equilibrium condition $C = T$, where $C = bxf_c = 200 \times 26x = 5220x$ and $T = 684\,000$ N assuming all safety factors equal to 1; thus, $x = 131$ mm. With x known, the lever arm of the couple of the longitudinal internal actions $C = T$ is $z = d - x/2 = 950 - 131/2 = 884$ mm. The condition $Ra_v = Tz$, where R is the reaction ($R = P/2$, where P is the total applied load), yields the value $R = V_1 = (z/a_v)T = (0.884/1.0) \times 684\,000 = 605$ kN which corresponds to failure of the 'horizontal' element of the 'frame' of the beam model.

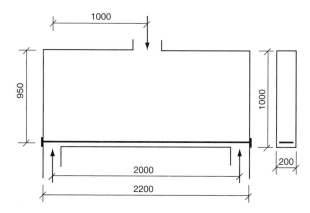

Fig. 4.23. Design details and position of point load for the beam with type IV behaviour

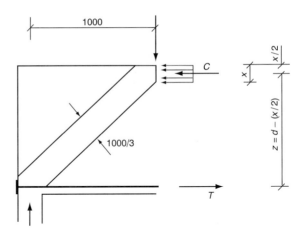

Fig. 4.24. Physical model of the beam in Fig. 4.23 and internal actions at the mid cross-section

On the other hand, the compressive strength of the inclined element of the 'frame' is exhausted when the axial force acting on it attains a value $C_\phi = b(a/3)f_c = 200 \times (1000/3) \times 26 = 1740$ kN. The vertical component V_2 of C_ϕ (i.e. the value transferred to the support should C_ϕ attain the value of 1740 kN) is $V_2 = C_\phi \sin\phi = 1740 \times 0\cdot663 = 1154$ kN. (Note that $\tan\phi = z/(L/2)$.) Naturally, the maximum load sustained by the beam should be the one which corresponds to the smallest of the two calculated values (V_1 and V_2) of the shear force V, i.e. $V = \min(V_1, V_2) = V_1 = 605$ kN, and, hence, $P = 2V = 1210$ kN. In fact, available experimental evidence[4.7] proved that the beam is indeed characterised by a flexural mode of failure, with the predicted value of the load-carrying capacity being equal to approximately 87% of the experimentally established value.

4.3. Simply-supported prestressed concrete beam

4.3.1. Physical model

The physical model depicted in Fig. 4.1, suitably modified, may also be used to describe the physical state of a simply-supported prestressed concrete (PSC) beam at its ultimate limit state. The modification is required in order to describe the effect of prestressing on the crack pattern of the beam and, hence, on the shape of the 'frame' which models the uncracked portion of the beam.

The modified physical model of a simply-supported prestressed concrete beam at its ultimate limit state is depicted in Fig. 4.25. This model differs from that of Fig. 4.1 only in respect of the position of the 'joint' of the horizontal and the inclined elements of the 'frame' of the proposed model. The location of this joint may be obtained from the following expression:

$$h = (d - x_g)(P_e/R) \quad (4.5)$$

where the geometric characteristics h, d, and x_g and the forces P_e and R are defined in Fig. 4.25. Clearly, the joint location is

DESIGN METHODOLOGY

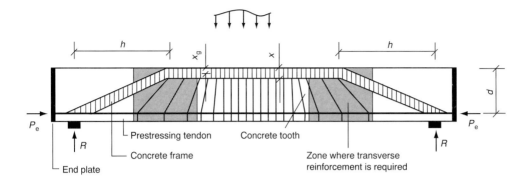

h = horizontal projection of the inclined portion of the CFP
P_e = effective prestressing force
R = reaction at support
d = effective depth of the member
x_g = centroidal distance of uncracked concrete from the compression face

Fig. 4.25. Physical model of a prestressed concrete beam

determined by the intersection of the resultant force at the support (i.e. vertical reaction plus prestressing force) with the level of the centroid of the uncracked portion of the beam.

It should be noted that, for the case of the two-point loading, the effect of prestressing is to shift the 'joint' away from the support. However, this location will always be situated within the shear span since, even under service conditions (i.e. within the elastic range of behaviour), the location of the 'joint' is at a_v. This follows from the shape of the bending-moment diagram for two-point loading, which is equivalent to a 'frame' with its joint located on the cross-section defining the shear span: since, in the elastic range, the tensile (T) and compressive (C) forces equal P_e, any moment increase in the flexural span is taken up by an increase in z (i.e. the 'height' of the 'frame') up to the stage when the distribution of stresses in the compressive zone departs from the triangular (linear) shape and the PSC beam begins to behave as an RC girder approaching ultimate conditions. It is at this stage that the z in the flexural span becomes practically constant, any further moment increase there being taken up by an increase in C and T, whereas in the shear span z can continue to increase up to the level of the horizontal member of the 'frame' which means that the joint of the 'frame' begins to shift towards the support. Thus, as the vertical load increases beyond the service conditions, the 'joint' position moves towards the support so that, at the ultimate limit state, the larger reaction corresponding to this increased external load results in a steeper inclined 'leg' that has now transposed the 'joint' to its closest position with respect to the support (see Fig. 4.26(a) and expression (4.5) corresponding to the limiting position of the joint at the ultimate limit state). The same reasoning can be applied to the cases where more than two point

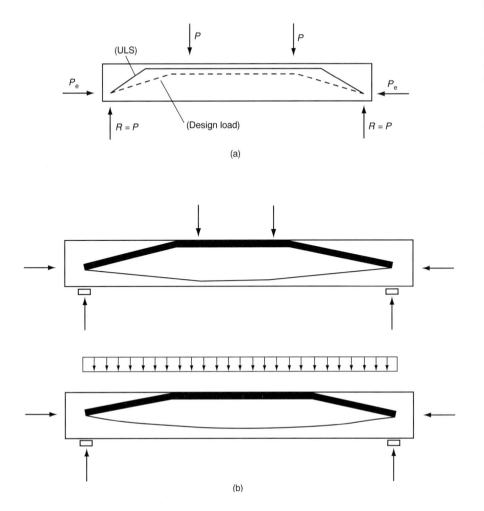

Fig. 4.26. (a) slopes of inclined leg at design load and ultimate limit state; (b) effect of tendon-profile shape on the physical model of a prestressed concrete beam

loads act on the beam, including the limiting instance of uniformly-distributed loading, with the position of the 'joint' defined by h. Thus, unlike the case of ordinary reinforced concrete members, where the 'joint' location could be either $a = 2d$ or $a = a_v$ depending on whether type II or type III behaviour, respectively, applied, in the case of prestressed concrete members $a = h$ always. With regard to the calculation of the 'critical' M_c, expression (4.1) is still valid, with $s = a_v$ for two-point loading but, in the case of more than two point loads/uniformly-distributed loading, h now replaces $2d$ so that $s = h$.

It should also be noted that, although the model of Fig. 4.25 refers to the case of a linear tendon profile, the model may be readily modified so that its use may be extended for any shape of tendon profile once the resultant of the external forces P_e and R are combined. Typical examples of tendon profiles and the corresponding prestressed concrete beam models are depicted in Fig. 4.26(b).

DESIGN METHODOLOGY

4.3.2. Failure criteria

A comparison of the physical models depicted in Figs 4.1 and 4.25 indicates that there is no qualitative difference between RC and PSC beam behaviour at the ultimate limit state. As a result, the diagram of Fig. 3.9, which describes the relation between load-carrying capacity and the geometric characteristics of RC beams without transverse reinforcement, may also be assumed to describe the behaviour of PSC beams without stirrups. However, as in most practical situations a PSC beam has a span significantly larger than that of an RC beam, with the larger portion of the applied load acting at a relatively large distance from the supports, the beam behaviour is usually of type I or II.

To this end, and as for the case of reinforced concrete, the load-carrying capacity of a PSC beam may be calculated *(a)* from the flexural capacity of the 'critical' beam cross-section, for the case of type I behaviour, and *(b)* from expression (4.1), which describes analytically portion II of the diagram of Fig. 3.9, for the case of type II behaviour. An extensive verification of the validity of the predictions of expression (4.1) for the case of PSC beams has been the subject of recently published research work.[4.8]

However, owing to the prestressing, the compressive stresses which develop within the end regions of the beam, i.e. within the inclined elements of the 'frame' of the model, are significantly larger than those developing in the corresponding regions of RC beams. Such stresses may lead to failure before the flexural capacity of the beams is exhausted. Such premature failure may be prevented by checking the compressive strength of the inclined elements of the 'frame' as described in Fig. 4.4 for the case of type IV behaviour (see section 4.2.2), but subject to the change proposed in section 4.3.4(e) below.

4.3.3. Assessment of reinforcement

The assessment of longitudinal reinforcement may be made as described in the literature referring to PSC structures,[4.9, 4.10] while the transverse reinforcement which may be required for the case of type II behaviour may be assessed, with the slight modification described below, as already described in section 4.2.5 for the case of RC beams. The modification involves the calculation of the shear force which acts at the beam cross-section through the 'joint' of the horizontal and inclined members of the 'frame' for the case of uniformly-distributed loading. As the joint lies at a distance h from the support (instead of $2d$, as for RC members), which is calculated from expression (4.5), the shear force is now given by $V_c = M_c(L-2h)/[h(L-h)]$, obtained by replacing $2d$ by h in its counterpart in section 4.2.5.

4.3.4 Procedure for checking shear capacity

The design of a PSC beam for flexural capacity (i.e. selecting suitable materials and cross-sectional characteristics, and assessing the longitudinal reinforcement and prestressing force required for the beam to have a given flexural capacity and structural performance under service loading conditions) is carried out as described in the pertinent bibliography.[4.9–4.11] In what follows, a procedure for checking shear capacity is proposed, aiming to prevent any type of failure before the flexural capacity of the beam is exhausted. This procedure comprises the following steps.

(a) *Construction of physical model.* As for the case of RC beams (see Fig. 4.1), it is evident from Fig. 4.25 that the shape of the physical model of the PSC beam is given. What is required is the position of the connection between the horizontal and inclined members of the 'frame' of the model. This position can easily be defined by using expression (4.2) so as to determine the centroid of the compressive zone, x_g, which defines the position of the horizontal member.

(b) *Assessment of flexural capacity.* At the ultimate limit state, there is not any qualitative difference in behaviour between RC and PSC beams. As a result, for given geometric characteristics, longitudinal reinforcement, and material quality, the assessment of flexural capacity (see references 4.9–4.11) essentially involves the calculation of the depth, x, of the horizontal element of the 'frame' of the physical model shown in Fig. 4.25. (It should be noted that the assessment of flexural capacity uses the simplified compressive stress block proposed in Fig. 4.13.) With x known, the lever arm of the internal longitudinal actions is obtained from $z = d - (x/2)$, and the flexural capacity is given by $M_f = Tz$.

(c) *Assessment of design shear forces.* The design shear forces are those corresponding to the flexural capacity of the beam.

(d) *Assessment of transverse reinforcement.* Transverse reinforcement may be required mainly in the region of the joint of the horizontal and inclined elements of the 'frame.' For the case of point loading, additional transverse reinforcement may also be required within the compressive zone in the region of such point loading to counter possible bond loss. Such reinforcement is assessed as described for the case of RC beams in section 4.2.5.

(e) *Checking of load-carrying capacity of inclined leg of 'frame' of model.* The load-carrying capacity of the inclined leg, which is subjected essentially to the resultant of the effective prestressing force and the reaction acting in the direction of the leg, may be calculated as described in section 4.2.2 except that, whereas for deep beams

DESIGN METHODOLOGY

$a \leq d$, for PSC members a usually exceeds d so that in checking the strength of the inclined leg in PSC beams the depth of this inclined leg should be limited to $d/3$ instead of $a/3$; such an approach is approximate (especially when the inclined leg of the 'frame' crosses the rectangular end block in T-beams) and, while appearing to be conservative (see subsequent example), further research is needed to refine this rough guideline. If the acting force is found to be larger than the calculated strength value, the width of the beam should be increased so that the fundamental condition *applied action $\leq$ strength* is satisfied.

4.3.5. Example of shear-capacity checking

Figure 4.27 shows the geometric characteristics and the reinforcement details of a PSC beam which has been one of the specimens whose behaviour was investigated in a research programme concerned with the verification of the validity of the proposed design method.[4.8] The figure also shows the loading arrangement used in the research programme. The beam was constructed with a concrete with $f_c = 44$ MPa, tendons with a maximum sustained stress $f_{pu} = 1900$ MPa and total cross-sectional area $A_p = 205 \cdot 4$ mm^2, and transverse reinforcement comprising bars with dia.1·5 and having a yield stress $f_{yv} = 460$ MPa. The long-term prestressing force was $P_e = 236 \cdot 49$ kN, while all safety factors were taken equal to 1.

Flexural capacity
The equation $C = A_c f_c = T$, where A_c is the cross-sectional area of the compressive zone and $T = A_p f_{pu} = 390\,260$ N, yields $A_c = 390\,260/44 = 8870$ mm^2 and $x = A_c/b = 8870/200 = 44$ mm. Hence, the lever arm of the internal longitudinal forces is $z = d - (x/2) = 218$ mm, while $M_f = Tz = 390\,260 \times 218 = 85 \times 10^6$ Nmm = 85 kNm (see Fig. 4.28).

Diagram of design shear forces
Equating the bending moments, with respect to the mid cross-section of the beam, of the applied point loads and the reaction at the left-hand side of the beam with the beam's flexural capacity yields the value of each of the point loads (corresponding to flexural capacity) as $P = 15 \cdot 62$ kN, with the reaction being $R = 3P = 46 \cdot 86$ kN. With these values of P and R, the shear-force diagram is as indicated in Fig. 4.29. From this diagram, the design shear force, within the portion of the beam between the support and the point load closest to it, is $V_f = 46 \cdot 86$ kN.

Physical model
Although the beam is subjected to six-point loading, the present case is treated as two-point loading since, as indicated in Fig. 4.30, the six loads are applied in the middle portion of the beam,

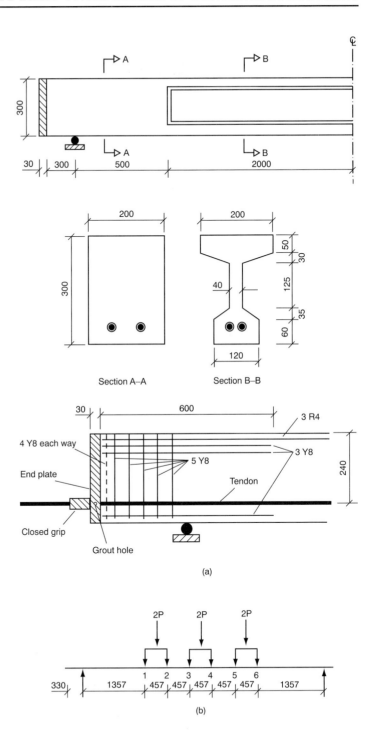

Fig. 4.27. PSC beam under multiple-point loading: (a) cross-section design details, and (b) loading arrangement

symmetrically about the mid-span cross-section, with the remainder of the beam comprising two large shear spans, each of length $a_v = 1357$ mm. The location of the joint of the horizontal and inclined elements of the 'frame' is found from expression

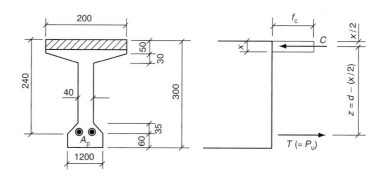

Fig. 4.28. Assessment of flexural capacity of the beam in Fig. 4.27

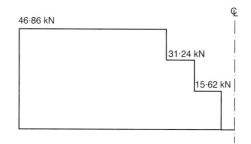

Fig. 4.29. Shear-force diagram of the beam in Fig. 4.27

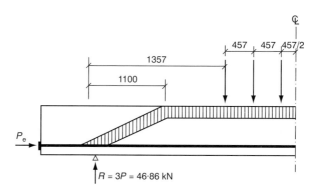

Fig. 4.30. Physical model of the beam in Fig. 4.27

(4.5) to be $h = 1100$ mm. As $h < a_v$, the beam should be characterised by either type I or type II behaviour depending on whether $M_c > M_f$ or $M_c \leq M_f$, respectively, with M_c being calculated from equation (4.1) and M_f being the flexural capacity.

Assessment of transverse reinforcement
With $s = a_v = 1357$ mm (two-point loading), equation (4.1) yields $M_c = 48.25$ kNm ($< M_f$), from which the shear force that can be sustained by concrete alone at the cross-section through the point load closest to the support is obtained from $V_c = M_c/s = 48.25/1.357 = 35.56$ kN $< V_f = 46.86$ kN. Hence, the transverse reinforcement (A_{sv}) required to sustain the shear force $V_f - V_c = 11.3$ kN, i.e. $A_{sv} = (V_f - V_c)/f_{yv} = 11\,300/460 = 24.56$ mm². The need for such an amount of reinforcement is satisfied by using 8 two-legged dia. 1·5 at 34 cc over a length $d = 240$ mm, symmetrically placed

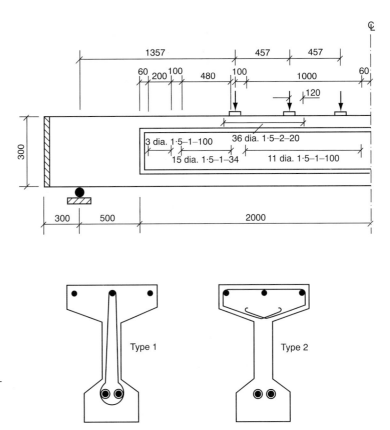

Fig. 4.31. Transverse-reinforcement details for the beam in Fig. 4.27

about the cross-section through the joint of the horizontal and inclined elements of the 'frame'; the beam in reference 4.8, however, was designed while the present design methodology was evolving and, hence, Fig. 4.31 shows that this reinforcement around the joint was extended to twice the actual distance required.

Owing to the point loading, transverse reinforcement may also be required in the compressive zone of the shear span of the beam in the region of the point load, in order to sustain the internal tensile actions which develop in this region when the bond between concrete and the longitudinal steel bars is destroyed (see section 3.2.5). Following the procedure described in section 4.2.5 and, by considering the equilibrium of the beam element between two consecutive flexural or inclined cracks for which there is no bond between the concrete and the longitudinal reinforcement (see Fig. 4.15), the amount of transverse reinforcement required results from the assessment of the following parameters:

- the increase Δz of the lever arm z of the internal longitudinal actions at the right-hand side of the element is $\Delta z = (V_\mathrm{f} - V_\mathrm{c})x/(2T) = (11\cdot3 \times 44)/(2 \times 390\cdot26) \approx 1$ mm;

- the depth of the compressive zone in the right-hand side of the element is $x_1 = 2(d-z-\Delta z) = 2 \times (240-218-1) = 42$ mm;
- the nominal compressive stress at the right-hand side of the element is $\sigma_1 = C/(bx_1) = 390260/(200 \times 42) = 46.46$ MPa > 44 MPa;
- the transverse nominal confining pressure (σ_{conf}), at the right-hand side of the element, which allows $\sigma_1 > f_c$, is $\sigma_{conf} = (\sigma_1 - f_c)/5 = (46.46-44)/5 = 0.49$ MPa;
- the transverse nominal tensile stress at the left-hand side of the element is numerically equal to σ_{conf}, i.e. $|\sigma_t| = |\sigma_{conf}|$;
- the transverse reinforcement required to sustain σ_t within a unit length of 100 mm is $A_{sv} = 100\sigma_t b/f_{yv} = 0.46 \times 100 \times 200/460 = 21.3$ mm^2/dm.

Links dia. 1·5–2–20 cc (this time, with the horizontal legs of the hoop stirrups continuous, see detail in Fig. 4.31) should be placed in the compressive zone within a length of the order of 240 mm adjacent to the left-hand side of the point load closest to the support (although, in the actual beam, which was designed before the present methodology was finalized,[4.8] this length was shorter), as well as extending it to the right of the point load, as indicated in Fig. 4.31, in order to allow for the loss of bond throughout the length where the shear force is considerable (i.e. beyond the second point load from the support).

Throughout the rest of the beam, nominal stirrups (two-legged, dia. 1·5 at 100cc) were provided capable of taking tensile stresses in the 40 mm web of the order of 0·5 MPa. This is a lower figure than the nominal 1 MPa adopted subsequent to the work in reference 4.8, in an attempt to make the design methodology more conservative.

The experimental results provided in reference 4.8 indicate that the transverse reinforcement designed by using the above method proved capable of preventing any type of failure occurring before the attainment of flexural failure. The load-carrying capacity established experimentally, $6 \times 15.41 = 92.5$ kN, was found to be practically equal to the predicted value of $6 \times 15.62 = 93.72$ kN.

Checking of load-carrying capacity of inclined leg of 'frame' of model

The acting force on the inclined leg is $C_a = \sqrt{P_e^2 + R^2} = \sqrt{55927.52 + 2195.86} = 241$ kN, whereas the maximum sustained force, calculated as described in sections 4.2.2, 4.3.2 and 4.3.4 is $C_\phi = b_w(d/3)f_c = 40 \times (240/3) \times 44 = 140.8$ kN. Although there is an apparent need to increase the width of the web, the fact that the beam attained the full flexural capacity shows that the present simplified criterion for checking the strength of the inclined leg of the 'frame' can be conservative (especially in the presence of a rectangular end block). However, in the absence of sufficient experimental data, it is recommended to use the proposed criterion which, in a design situation, would have resulted in an increase of b_w to about 70 mm.

4.4. Skeletal structural forms with beam-like elements
4.4.1. Physical models

The application of the design methodology presented in the preceding sections can be extended not only to structural-concrete members other than simply-supported beams but, also, to more complex structural configurations comprising beam-like elements. Examples of the use of the model in Fig. 4.1 for the case of various types of structural-concrete members are illustrated in Figs 4.32 and 4.33, while Fig. 4.34 presents a characteristic portion of a more complex structural configuration whose modelling is also based on the model of Fig. 4.1.

Figure 4.32 shows that a cantilever, subjected to a point load near its free end, may be designed as a simply-supported beam under a single point load applied at its mid cross-section, since the boundary conditions at the fixed end of the cantilever are similar to conditions at the mid-span of a simply-supported beam. Similarly, a structural concrete wall under horizontal loading, being essentially a cantilever beam, may also be designed by using the proposed methodology. In fact, the application of the proposed methodology to the design of structural concrete walls was found to yield safe and efficient design solutions,[4.12] in spite of the considerably smaller amount of transverse reinforcement required in comparison with that specified by current codes.

Figure 4.33 depicts a reinforced concrete beam fixed at both ends, such as, for example, the beam coupling two structural walls. Such a beam, as for the case of the cantilever of Fig. 4.32, may also be designed by using the proposed methodology; the beam can be divided into two portions extending between the beam's fixed ends and the point of contraflexure, each of them essentially functioning as a cantilever. In this case, however, the design method must be complemented so as to allow for the design of the connection of the two 'cantilever' beams. The connection may be modelled as an 'internal support' effected within a length equal to the beam depth d and symmetrical about the point of contraflexure, where the 'free' end of the one cantilever supports (or is supported by) the 'free' end of the other cantilever. The 'internal reaction' which develops in such a support is equivalent to the shear force acting at the cross-section through the point of contraflexure.

Assuming that concrete is capable of sustaining the portion (V_c) of the 'internal reaction' (shear force) which corresponds to a nominal tensile stress of $f_t - 1$ MPa acting over the length d of the 'internal support' (i.e. $V_c = f_t bd$), placing links in sufficient quantity to sustain the remainder of the shear force acting at such an 'internal support' has been found to yield a satisfactory design

DESIGN METHODOLOGY

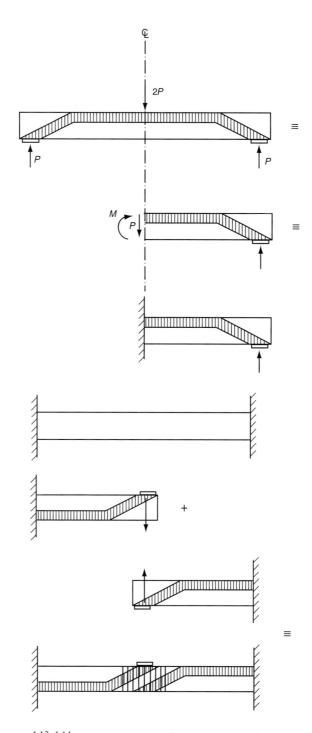

Fig. 4.32. Use of physical model of a simply-supported beam for modelling a cantilever

Fig. 4.33. Use of physical model of a cantilever for modelling a fixed-end beam

solution.[4.13, 4.14] Such links are placed within a length equal to twice the beam depth ($2d$), symmetrically located about the point of contraflexure (thus doubling the extent of the region of contraflexure and the total amount of transverse reinforcement

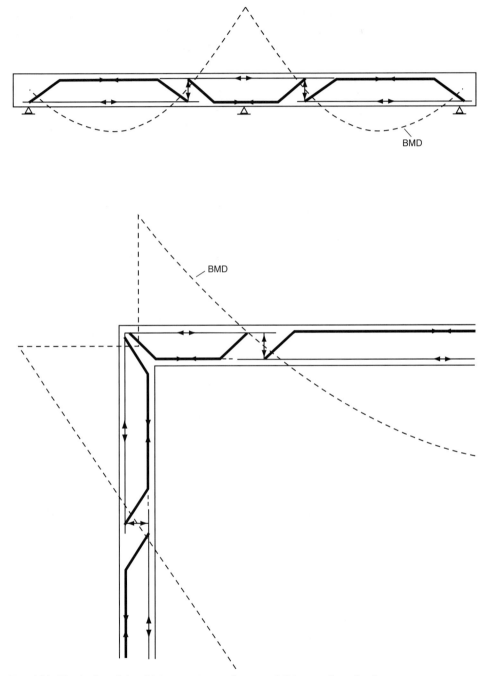

Fig. 4.34. Physical models of (a) a continuous beam and (b) a portion of a frame

through such an in-built safety measure). If concrete is capable of sustaining the full shear force, a nominal number of links is placed, within the same length (i.e. $2d$), capable of sustaining tensile stresses of 1 MPa.

DESIGN METHODOLOGY

The method described above for the application of the proposed design methodology is indicative of the manner in which this methodology may be applied to the design of any structural form comprising beam-like members. As indicated in Fig. 4.34, any frame-like structural configuration may be divided into portions spanning between two consecutive points of contraflecture. Each of these portions may then be considered equivalent to a simply-supported beam, while the connections of these portions may be designed as 'internal supports', which has been already described for the instance of the beam with fully-fixed ends shown in Fig. 4.33.

4.4.2. Design procedure

The sizing of any structural concrete configuration with beam-like members may be carried out as described in section 4.2.6 for the case of a simply-supported beam. With the geometric characteristics given, structural analysis yields the internal actions in the form of bending-moment, axial-force, and shear-force diagrams. From these diagrams, the bending-moment diagram is used for locating the points of contraflexure, so as to identify the portions of the structure between sets of two consecutive such points: these portions of the span can then be modelled as for the case of the simply-supported beam depicted in Fig. 4.1. As described in the preceding section, the connection of two consecutive portions of the structure may be considered as an internal support which is effected with the provision of transverse reinforcement in a quantity either nominal, for the case where concrete is capable of sustaining the total 'internal reaction' (shear force) acting at the point of contraflexure, or sufficient to sustain the portion of the 'internal reaction' in excess of that that can be sustained by concrete alone.

4.4.3. Design examples

The design examples presented in the following have been taken from publications[4.13, 4.14] concerned with the verification of the validity of the proposed methodology. The design was based on the assumption that bond failure is unlikely to occur for the rectangular cross-sections considered (this is based on past experimental evidence, and it could also be argued that stress concentrations are less likely to occur in rectangular cross-sections than in T-shaped ones) and under the monotonic loading conditions imposed on the structural forms investigated: hence, no attempt was made to assess transverse reinforcement capable of sustaining the tensile stresses that develop in the compressive zone due to bond failure. The results obtained from the experimental investigation of the structural forms designed to the proposed method also verified the validity of the above assumption. However, in general (but especially for T-sections and for the case of loading histories involving large load reversals), it is considered unsafe to ignore the likelihood of bond failure.

(a) Simply-supported beam with overhang

Figure 4.35 presents the geometric characteristics of the beam, together with the reinforcement detailing and the loading arrangement.[4.14] The beam was constructed by using a concrete with $f_c = 30$ MPa, longitudinal reinforcement with yield and maximum stresses equal to $f_y = 600$ MPa and $f_{su} = 870$ MPa respectively, and transverse reinforcement with a yield stress of $f_{yv} = 240$ MPa. Using a value of 1 for all safety factors, the beam's flexural capacity was calculated by using the full stress–strain curve of the longitudinal steel bars, which was available in reference 4.14, and was found to be $M_f = 33.41$ kNm. For the loading arrangement indicated in Fig. 4.35, the above value of M_f yields a loading-carrying capacity (total load) of $4P = 133.64$ kN, leading to the bending-moment and shear-force diagrams indicated in Fig. 4.36. These diagrams were used to design the transverse reinforcement required to prevent any type of failure from occurring before the attainment of flexural capacity.

The physical model, which, in compliance with the procedure described in section 4.4.1, is compatible with the bending-moment diagram of Fig. 4.36(b), is depicted in Fig. 4.37. The figure indicates that the structure comprises two 'simply-supported' beams forming between consecutive points of 'zero bending moment' and interacting at the point of contraflexure (defined by the zero bending moment of the diagram in Fig. 4.36(b)) in the manner described in section 4.4.1. Each of the above 'simply-supported' beams (one of them, with no horizontal member to the frame model for reasons explained at the end of

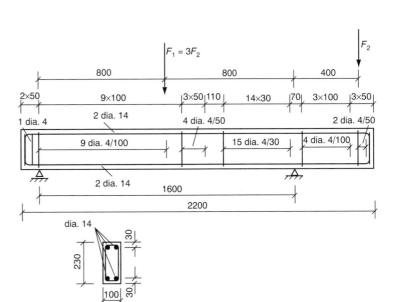

Fig. 4.35. Design details and loading arrangements for the simply-supported beam with an overhang

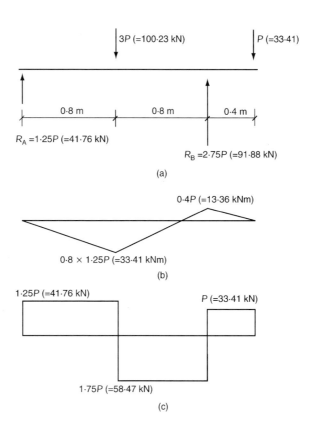

Fig. 4.36.
Arrangement and magnitude of loads and corresponding bending-moment and shear-force diagrams of the beam in Fig. 4.35: (a) expected load; (b) BMD; (c) SFD

section 4.2.2) is designed in the manner described in section 4.2, with the interaction being effected as described in section 4.4.1. Premature failure (i.e. failure before the flexural capacity is exhausted) of the physical model may occur, on the one hand, in the region of the joints of the horizontal and inclined elements of the 'frame' of the model for each of the two 'simply-supported' beams (i.e. locations 1 to 4), and, on the other hand, in the region of the 'internal support' (i.e. location 5), where the interaction of the two beams is effected (see Fig. 4.37).

The left-hand side 'simply-supported' beam is subjected to point loading with shear spans $a_{v1} = 800$ mm $> 2d$ ($= 2 \times 200 = 400$ mm) and $a_{v2} = 570$ mm $> 2d$ ($= 2 \times 200 = 400$ mm), thus both being characterised by type II behaviour. For $s = a_{v1} = 800$ mm and $s = a_{v2} = 570$ mm, expression (4.1) gives $M_{c1} = 24.125$ kNm and $M_{c2} = 19.642$ kNm, respectively, and, hence, the values of the shear force that can be sustained by concrete alone at locations 1 and 2 are $V_{c1} = 24.125/0.8 = 30.16$ kN and $V_{c2} = 19.642/0.57 = 34.46$ kN, with both values of the shear force being smaller than their design counterparts $V_{d1} = 41.76$ kN and $V_{d2} = 58.47$ kN (see Fig. 4.36 (c)). The excess amounts of shear force $V_{s1} = 41.76 - 30.16 = 11.6$ kN and $V_{s2} = 58.47 - 34.46 = 24.01$ kN are sustained by transverse

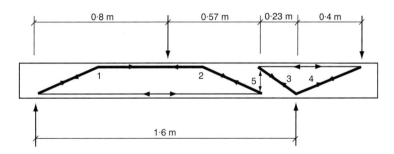

Fig. 4.37. Physical model of the beam in Fig. 4.35

reinforcement which, for $f_{yv} = 240$ MPa, is $A_{sv1} = 11 \cdot 6 \times 10^3/240 = 48 \cdot 33$ mm^2 (2 dia. 4 two-legged stirrups) and $A_{sv2} = 24 \cdot 01 \times 10^3/240 = 100 \cdot 04$ mm^2 (4 dia. 4 two-legged stirrups), placed within a length of $d = 200$ mm symmetrically about locations 1 and 2, respectively.

The right-hand side 'simply-supported' beam is also subjected to point loading, the point load being the reaction at the right-hand support of the 'real' beam, while the supports are the point of application of the external load acting at the overhang and the 'internal reaction' at location 5. The shear spans of this beam are $a_{v3} = 230$ mm $< 2d$ ($= 400$ mm) and $a_{v4} = 400$ mm $= 2d$ ($= 400$ mm), both being characterised by type III behaviour. For location 3, $M_{c3} = 30 \cdot 787$ kNm (corresponding to $a_v/d = 1 \cdot 15$) is obtained by linear interpolation between $M_f = 33 \cdot 41$ kNm (corresponding to $a_v/d = 1$) and $M_c = 15 \cdot 923$ kNm (corresponding to $a_v/d = 2$), while for location 4, $M_{c4} = 15 \cdot 923$ kNm resulting from expression (4.1) for $s = 2d = 400$ mm. Both M_{c3} and M_{c4} are larger than their design counterparts, $M_{d3} = M_{d4} = 13 \cdot 36$ kNm, respectively, indicated in Fig. 4.36(b), and, hence, only nominal reinforcement is required which is assessed so as to be capable of sustaining a tensile stress of 1 MPa per unit length of the beam, i.e. for $f_{yv} = 240$ MPa, $A_{sv, \text{nominal}} = 1000 \times 100 \times 1/240 = 416 \cdot 67$ mm^2/m (= dia. 4 two-legged stirrups at 60 mm cc) which is more than the amount provided (dia. 4 two-legged stirrups at 100 mm cc) and found sufficient for preventing brittle failure. (The amount of nominal reinforcement provided was calculated so as to sustain a tensile stress of 0·5 MPa instead of the value of 1 MPa recommended in the present book.) A similar amount of nominal reinforcement is also placed throughout the remainder of the beam where the proposed model does not require any checking of the shear capacity.

At location 5 ('internal support' developing over a length equal to $d = 200$ mm symmetrically about the point of contraflexure), the portion of the 'internal reaction' sustained by concrete alone (V_{c5}) is assessed by assuming that the tensile strength of concrete is 1 MPa. Then, $V_{c5} = 1 \times 200 \times 100 = 20$ kN, which, as indicated in the diagram of the shear forces in Fig.

4.36(c), is smaller than the applied shear force $V_{d5} = 58\cdot47$ kN, and, hence, transverse reinforcement A_{sv5} is required in order to sustain the portion $V_{s5} = V_{d5} - V_{c5} = 38\cdot47$ kN, in excess of that that can be sustained by concrete alone. For $f_{yv} = 240$ MPa, $A_{sv} = V_s/f_{yv} = 10^3 \times 38\cdot47/240 = 160.3$ mm^2 within a length $d = 200$ mm (i.e. 7 dia. 4 two-legged stirrups). Such an amount of reinforcement is placed over a length of 200 mm on both sides of location 5.

The details of the transverse reinforcement assessed by using the proposed methodology are shown in Fig. 4.35. The experimental verification of the validity of the proposed design methodology indicated that, although the amount of transverse reinforcement placed was significantly smaller than that required by current codes (see section 1.6), the beam eventually failed, as predicted, in flexure.[4.14]

(b) **Cantilever**

Figure 4.38 provides the design details together with the testing arrangement which induced a combined action of shear force and bending moment at the 'free' end of the cantilever.[4.14] The materials used for the construction of the cantilever were those also used for the beam discussed in the preceding section. It may be noted in Figs 4.35 and 4.38 that both the simply-supported beam and the cantilever had identical cross-sectional characteristics and, hence, the flexural capacity of the cantilever is that of the beam, i.e. $M_f = 33\cdot41$ kNm.

Figure 4.39 shows the combination of the applied loads which causes flexural failure together with the corresponding bending-moment and shear-force diagrams. The physical model of the structure, which, in compliance with the methodology discussed in section 4.4.2, is compatible with the bending-moment diagram of Fig. 4.39, is shown in Fig. 4.40. The figure indicates that the structure comprises two 'cantilevers' extending between the top and bottom ends of the structure and the point of contraflexure (defined by the zero bending moment of the diagram in Fig. 4.39(b)), where they interact in the manner described in section 4.4.2. Each of the 'cantilevers' is modelled as a 'half simply-supported beam' in the manner described in section 4.4.1 (see also Fig. 4.32), while the interaction of the cantilevers is effected by the provision of transverse reinforcement, the function of which, as indicated in Fig. 4.33, resembles that of an internal support. (Recall the single-member instance in the frame model when $a_v/d \leq 2$, which presently is the case ($a_v/d = 2$) for the upper 'cantilever'.)

It appears from the physical model in Fig. 4.40 that failure, other than flexural, may occur either at the locations of the joints of the horizontal and inclined elements of the 'frames' of the model (i.e. locations 1 and 2) or in the region of the 'internal

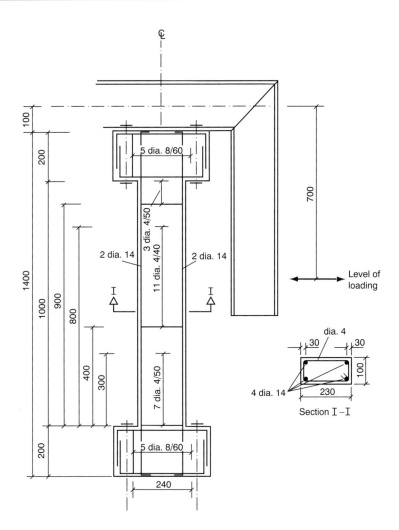

Fig. 4.38. Design details and loading arrangement of the cantilever

support' (i.e. location 3). As $a_v/d = 600/200 = 3 > 2$ (i.e. type II behaviour) for the lower of the two 'cantilevers' comprising the structure, the shear force (V_c) that can be sustained by concrete alone at location 1 may be calculated by using expression (4.1) which, for $s = a_v = 600$ mm, yields $M_c = 20.376$ kNm and, hence, $V_c = M_c/s = 20.376/0.6 = 33.96$ kN. The transverse reinforcement required in location 1 to sustain the shear force $V_s = V_d - V_c$ $= 55.68 - 33.96 = 21.72$ kN, in excess of that that can be sustained by concrete alone in this location, is $A_{sv} = V_s/f_{yv} = 21\,720/240 = 90.5$ mm^2. Such reinforcement should be placed over a length $d = 200$ mm spread symmetrically about the joint in location 1. In fact, Fig. 4.38 indicates that the transverse reinforcement placed in the above region comprises four two-legged stirrups, each with a diameter of 4mm, i.e. 4 dia. 4 = 100.53 mm^2, which is slightly larger than the amount required.

DESIGN METHODOLOGY

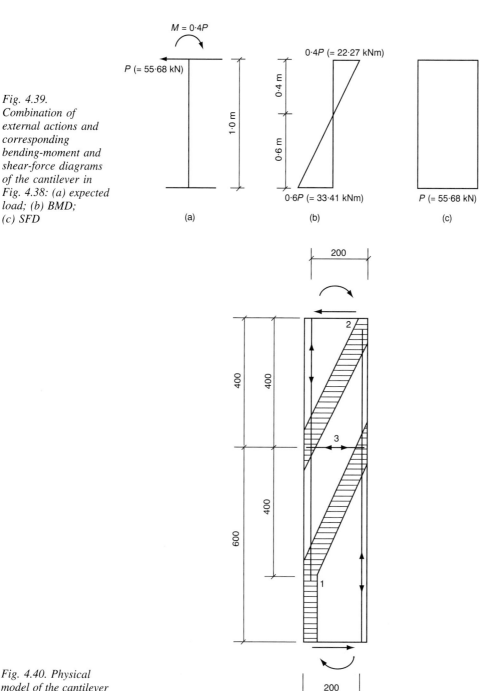

Fig. 4.39. Combination of external actions and corresponding bending-moment and shear-force diagrams of the cantilever in Fig. 4.38: (a) expected load; (b) BMD; (c) SFD

Fig. 4.40. Physical model of the cantilever in Fig. 4.38

With regard to the upper of the two 'cantilevers' comprising the structure, $a_v/d = 400/200 = 2$ and, hence, the 'cantilever' is characterised by type III behaviour (see section 4.2.2 and Figs 4.1 and 4.2). In the absence of transverse reinforcement, the bending moment M_c that can be sustained for such type of behaviour in location 2, can be obtained directly from expression (4.1) since $a_v/d = 2$ (see section 4.2.2). Equation (4.1) yields $M_c = 16 \cdot 02$ kNm which is smaller than its design counterpart $M_d = 22 \cdot 27$ kNm (see Fig. 4.39(b)), and, hence, transverse reinforcement is required in order to increase the flexural capacity in location 2 by an amount $\Delta M = M_d - M_c$. As described in section 4.2.5, such reinforcement should be uniformly distributed throughout the shear span a_v (which, in the present case, is equal to the length of the upper 'cantilever' of the structure) and be capable of sustaining a total tensile force $T_{yv} = 2\Delta M/a_v = 2(22 \cdot 27 - 16 \cdot 02)/0 \cdot 4 = 31 \cdot 25$ kN. Hence, the amount of reinforcement required is $A_{sv} = T_{yv}/f_{yv} = 31\,250/240 = 130 \cdot 21$ mm^2 (i.e. 6 dia. 4 two-legged stirrups) placed throughout the 400 mm length of the upper 'cantilever' of the structure. In fact, Fig. 4.38 indicates that the amount of transverse reinforcement placed in the above region comprises 7 dia. 4 two-legged stirrups.

Assuming that the tensile strength of concrete is 1 MPa, the shear force, sustained by concrete within a length $d = 200$ mm in the region of location 3, is $V_c = 1 \times 200 \times 100 = 20$ kN, which, as indicated in the diagram of the shear forces in Fig. 4.39(c), is smaller than the applied shear force $V_d = 55 \cdot 68$ kN, and, hence, transverse reinforcement A_{sv} is required in order to sustain the portion $V_s = V_d - V_c = 35 \cdot 68$ kN, in excess of that that can be sustained by concrete alone. For $f_{yv} = 240$ MPa, $A_{sv} = V_s/f_{yv} = 10^3 \times 35 \cdot 68/240 = 148 \cdot 67$ mm^2 (i.e. 6 dia. 4 two-legged stirrups) within a length $d = 200$ mm. Such an amount of reinforcement is placed over a length of 200 mm on both sides of location 3, i.e. a total of 12 dia. 4 two-legged stirrups, which is slightly larger than the amount indicated in Fig. 4.38.

It may be noted in Fig. 4.38 that, although the regions (associated with locations 2 and 3) requiring reinforcement overlap, the amount of reinforcement placed in the overlap is not the sum of the amounts specified for locations 2 and 3 within the overlap, but the larger of these two values.

Finally, it is interesting to note that, as for the case of the beam discussed in the preceding section, the tests carried out on the cantilevers revealed that, in spite of the smaller amount of transverse reinforcement in comparison with that specified by current codes (see section 1.5), the cantilevers eventually failed, as predicted by the design method used, in flexure.[4.14]

(c) Continuous beam

Figure 4.41 provides the design details together with the testing arrangement which was used to test to destruction a two-span beam fully reported in the literature.[4.13] The materials used for the construction of this continuous beam were concrete with $f_c = 50$ MPa, longitudinal compression and tension steel bars

DESIGN METHODOLOGY

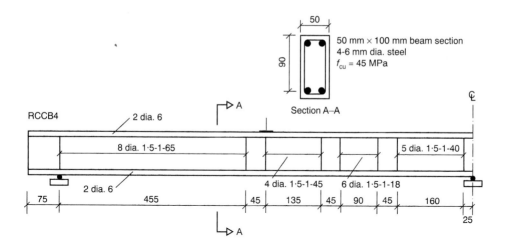

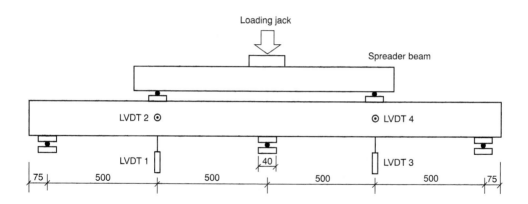

Fig. 4.41. Design details and loading arrangement of the continuous beam

(2 dia. 6) with cross-section $A_s = 56.55$ mm^2 and yield and ultimate stresses $f_y = 570$ MPa and $f_{su} = 665$ MPa respectively, and stirrups with yield and ultimate stresses $f_{yv} = 460$ MPa and $f_{suv} = 510$ MPa respectively. In what follows, the proposed design method is used to establish whether or not the amount of transverse reinforcement adopted is sufficient to prevent the occurrence of any type of brittle failure before flexural capacity is exhausted. (Note that the presence of compression reinforcement may, in principle, be allowed for in expression (4.1) by the change in lever arm z between the resultant tensile and compressive forces at the relevant cross-section due to this additional reinforcement.)

Using safety factors equal to 1, the procedure described in Fig. 4.42 yields the flexural capacity as $M_f = 3.05$ kNm. Assuming that collapse of the beam occurs owing to the formation of 'plastic' hinges, the elastic analysis of the beam indicates that 'plastic' hinges form at the locations of the internal support and

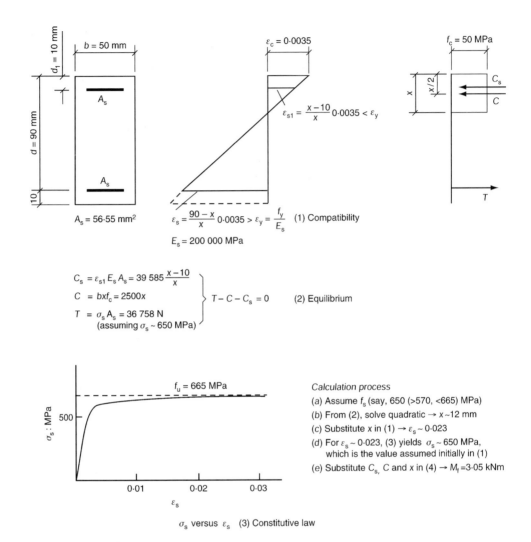

Fig. 4.42. Procedure for assessing the flexural capacity of the beam in Fig 4.41

at around the mid-spans of each of the two portions of the beam. For such a failure mechanism, in which plastic hinges are assumed to form at the internal support and at each of the two mid-spans, the classic virtual-work equation for the idealized plastic collapse mechanism yields $P = 6M_f$. The predicted load-carrying capacity (total applied load) of the beam, which corresponds to the above value of the flexural capacity, is, therefore, $2P_{pred} = 2 \times 18 \cdot 3 = 36 \cdot 6$ kN, while the value established experimentally is $2P_{exp} = 2 \times 21 \cdot 425 = 42 \cdot 85$ kN.

The deviation of the predicted value of the load-carrying capacity from the experimental one should not be attributed entirely to the method used for the calculation of flexural capacity. In fact, it is considered that the above deviation is due primarily to the development, at the free-sliding supports, of frictional forces which restrained the horizontal deformation of the beam in bending. Such a restraint, which is equivalent to the application of a relatively small axial compressive force, is capable of leading to an increase of the flexural capacity sufficient to justify the above deviation of the predicted value of the load-carrying capacity from its experimental counterpart.[1.10]

Assuming the formation of plastic hinges at the cross-section through the internal support and the mid cross-section of the two spans, the bending-moment and shear-force diagrams, which correspond at the above values of the load-carrying capacity, are depicted in Fig. 4.43 (the continuous and dashed lines describe the experimental and predicted values respectively). The figure also illustrates the physical model of the beam, which is compatible with the bending-moment diagram, in which locations 1 to 4 (with locations 2 and 4 being identical, as will be seen later), where a check of shear capacity is required in compliance with the proposed methodology, are clearly marked.

Figure 4.43(c) indicates that the structure is subdivided into three 'simply-supported beams' that form between consecutive locations of zero values of the bending moments and interact at the points of contraflexure (marked as location 3 in the figure). As for the case of the two design examples presented in the preceding sections, the interaction is considered to be effected by using transverse reinforcement which may be viewed to act as an 'internal support' (see section 4.4.2). As both the structure (continuous beam) and the applied load are symmetrical about the section through the interior support, the two outer 'simply-supported beams' comprising the structure are identical, and the middle 'simply-supported beam,' which is also symmetrical about the above section, is under the action of the reaction at the middle support of the continuous beam.

It may be noted in Fig. 4.43(c) that the length of the smaller shear span (i.e. the one which is closer to the central support) of the outer 'simply-supported beam' is equal to the length of the shear span of the middle symmetrical 'simply-supported beam'. Therefore, the 'frames' of the physical models of these two beams are identical within the above shear spans, namely the joints marked as locations 2 and 4 are identical.

Since for both shear spans including locations 1 and 2, $a_v/d > 2$, and they hence conform to type II behaviour, the tensile force that can be sustained by concrete at locations 1 to 2 may be obtained by $V_c = M_c/s$, where s is the value of the shear span corresponding to each of the above locations, and M_c is the bending moment calculated from expression (4.1) by using the values of s corresponding to locations 1 and 2 respectively. On the other hand, the tensile force that may be sustained by the stirrups is $V_s = A_{sv} f_{yv} n$, where A_{sv} and f_{yv} are the cross-section

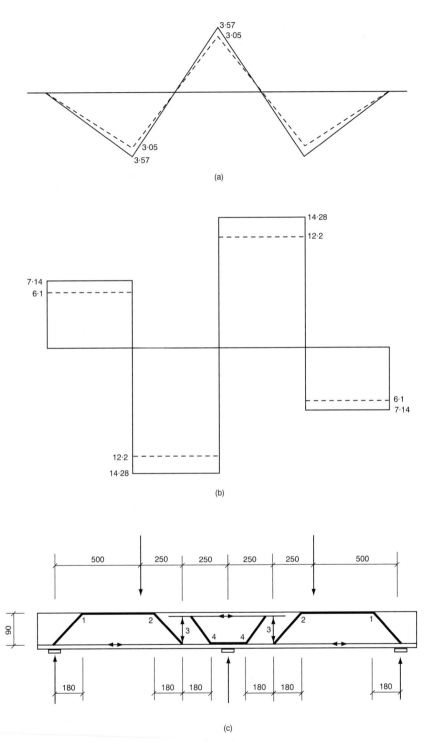

Fig. 4.43. (a) BMD, (b) SFD and (c) physical model of the continuous beam in Fig. 4.41 (note: in the BMD and SFD continuous and dashed lines denote values from plastic analysis and experiment respectively)

and yield stress of one two-legged stirrup respectively, and n is the total number of stirrups placed, within a length equal to the beam depth, symmetrically about the location at which shear capacity is checked. If s' is the stirrup spacing, then $n = d/s' + 1$.

In accordance with the above, the shear capacity at location 1 is $V_1 = V_{c1} + V_{s1} = M_{c1}/a_{v1} + A_{sv}f_{yv}n_1 = 2 \cdot 96/0 \cdot 5 + 2(\pi 1.5^2/4)\,460 \times 2 = 5 \cdot 92 + 3 \cdot 25 = 9 \cdot 17$ kN which, as indicated from the shear-force diagram of Fig. 4.43, is larger that the acting shear force found experimentally to be 7·14 kN. On the other hand, the shear capacity at location 2 is $V_2 = V_{c2} + V_{s2} = M_{c2}/a_{v2} + A_{sv}f_{yv}n_2 = 1 \cdot 92/0 \cdot 25 + 2(\pi 1 \cdot 5^2/4)\,460 \times 3 = 7 \cdot 68 + 4 \cdot 88 = 12 \cdot 56$ kN which is, again, larger than the design value of 12·2 kN. Although $V_2 = 12 \cdot 56$ kN was actually slightly smaller than the acting shear force found experimentally to be 14·28 kN, it did not lead to the failure of the beam. This small additional hidden margin of safety against a 'shear' type of failure may be attributed partly to the strain-hardening of the transverse reinforcement.

In compliance with the methodology described in section 4.4.1, concrete alone in the region of the 'internal support' is capable of sustaining a force V_{c3} corresponding to a nominal tensile stress of 1 MPa acting over the length d ($=90$ mm) of the internal support, i.e. $V_{c3} = 1 \times 50 \times 90 = 4 \cdot 5$ kN, while the contribution of the stirrups is $V_{s3} = 6 \times 2(\pi 1 \cdot 5^2/4)\,460 = 9750$ N $= 9 \cdot 75$ kN. Hence, the total force sustained at location 3 is $V_3 = 4 \cdot 5 + 9 \cdot 75 = 14 \cdot 75$ kN, which, as indicated by the shear-force diagram (see Fig. 4.43(b)), is sufficient to sustain the acting shear force at this location.

(d) **Portal frame with fixed ends**
Figure 4.44 shows the design details of a portal frame, designed in compliance with the proposed method. A similar portal frame was also designed to current code provisions[1.3] for purposes of comparison. The frame of Fig. 4.44 was one of the structural elements tested in an experimental research programme concerned with an investigation of the validity of the proposed methodology.[4.13] The frame was constructed by using concrete with $f_c = 43$ MPa, longitudinal steel bars (dia. 10) with yield stress $f_y = 560$ MPa and ultimate strength $f_{su} = 680$ MPa, and transverse reinforcement (stirrups) comprising wires (dia. 1·5) with yield stress $f_{yv} = 460$ MPa and ultimate strength $f_{suv} = 510$ MPa. It is important to note in the figure that the design details resulting from the use of the proposed method differ from those resulting from the use of the code[1.3] only in respect of the additional transverse reinforcement specified by the proposed method at locations marked as IS2 and IS3 in Fig. 4.45(c).

The frame was designed by using the bending-moment and shear-force diagrams depicted in Fig. 4.45, together with the resulting physical model of the frame, providing the basis for the

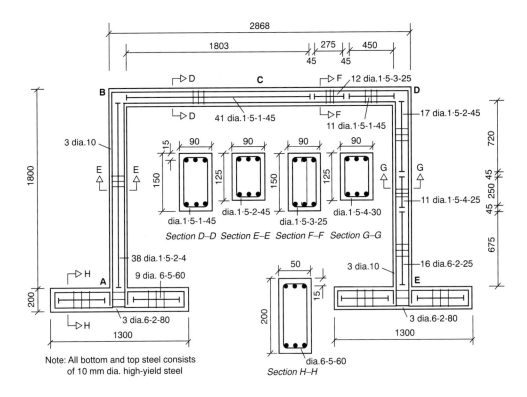

Fig. 4.44. Design details of the portal frame with fixed ends

proposed methodology. With regard to the corner joints of the portal frame, it should be noted that their design was based on conventional methods (see reference 4.13) and, hence, will not be discussed in what follows. The diagrams resulted from the elastic analysis of the structure under the combined action of a vertical load $V = 24$ kN, acting at mid-span of the horizontal member of the frame, and a horizontal load $H = 20$ kN, acting at the left-hand end of the same member along its longitudinal axis. From the design details depicted in Fig. 4.44 the flexural capacities of the horizontal and vertical members of the frame predicted by the proposed design method are found to be 16·05 kNm and 13·67 kNm respectively.

Figure 4.46 depicts the (sequential) loading history intended together with that actually achieved. The figure shows that the intended first stage of loading was that actually achieved, i.e. the frame was subjected to the action of V alone which increased to a value of 24 kN. During the second stage of loading the intention was for V to remain constant and equal to the value attained in the preceding loading stage, while H increased progressively to failure. However, it was established that, during this second stage, V continued to increase such that $\Delta V/\Delta H \approx 0.73$ (0·23, for the case of the portal frame designed to the code), where ΔV and ΔH are the increments of V and H respectively during this loading stage. The total values of V and H that eventually caused

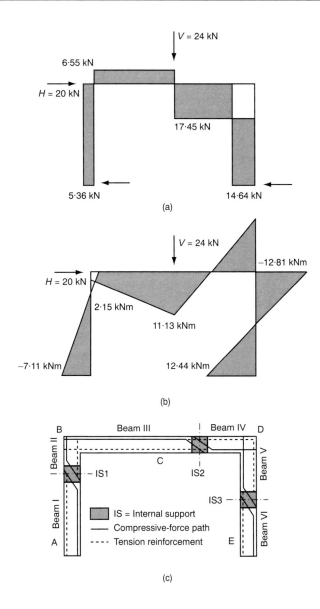

Fig. 4.45. (a) BMD (elastic), (b) SFD (elastic) and (c) physical model used to design the portal frame in Fig. 4.44

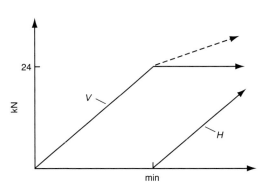

Fig. 4.46. Intended (continuous) and actual (dashed) loading histories for the portal frame in Fig. 4.44

failure of the frame designed by the proposed method were 38·68 kN and 19·95 kN, respectively, while the frame designed to the code failed under the combined action of $V = 28·5$ kN and $H = 20$ kN.

Considering the case of the frame designed by the proposed method, for the above values of V and H, the elastic analysis of the structure shows that the formation of plastic hinges at the ends of the right-hand side vertical member and at the mid-span of the horizontal member of the frame is inevitable. Assuming the formation of these three plastic hinges, a plastic analysis of the frame can be carried out, resulting in bending-moment and shear-force diagrams similar to the ones depicted in Fig. 4.47. The latter figure, however, shows the actual experimental internal actions at failure and also includes a schematic representation of the corresponding physical model of the frame which differs from the original one based on 'internal-support' locations corresponding to points of contraflexure obtained from elastic analysis (rather than plastic analysis or experiment). The corner joints, which, as mentioned earlier, were designed by conventional methods,[4.13] are treated as rigid bodies.

The adequacy of the transverse reinforcement provided may be checked as described in the following for the specific cases of locations 1, 2, and 3. As locations 1 and 2 lie within shear spans a_v larger than $2d = 270$ mm, i.e. $a_v/d > 2$, the shear force that can be sustained by concrete at these locations may be established by using expression (4.1) for the calculation of M_c. They were found to be $V_{c1} = M_{c1}/a_{v1} = 13·44/0·741 = 18·14$ kN and $V_{c2} = M_{c2}/a_{v2} = 12·19/0·631 = 19·32$ kN respectively, while the stirrups (at both locations) sustain a shear force $V_{s1} = V_{s2} = 2(\pi 1·5^2/4) 460 \times 4 = 6·50$ kN. The values of the total shear capacity, therefore, are $V_1 = 18·14 + 6·50 = 24·64$ kN and $V_2 = 19·32 + 6·50 = 25·82$ kN, against a value of 21·67 kN for the acting shear force. At location 3, concrete sustains a shear force $V_{c3} = 1 \times 135 \times 90 = 12·15$ kN, while the stirrups sustain a shear force $V_{s3} = 3 (\pi 1·5^2/4) 460 \times 6 = 14·63$ kN. Hence, the value of the total shear capacity at location 3 is $V_3 = 12·15 + 14·63 = 26·78$ kN, against a value of 21·67 kN for the acting shear force. As a result, there is an adequate margin of safety against any type of failure connected with the presence of a shear force at the locations which were checked. The same conclusions can be drawn by checking locations 4, 5, and 6 of the horizontal member, and locations 7 and 8 of the vertical member. These calculations vindicate the desirability of the amount and spread (over a length of $2d$ rather than d) of the internal-support transverse reinforcement so as to allow for possible shifts in contraflexure points.

The difference in behaviour exhibited by the two frames tested is apparent in Figs 4.48 and 4.49 which depict typical load–

DESIGN METHODOLOGY

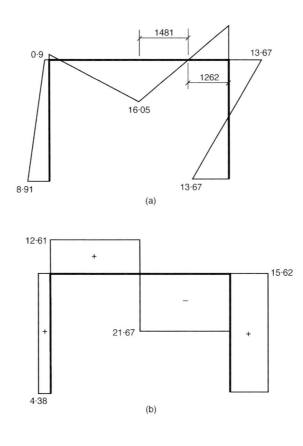

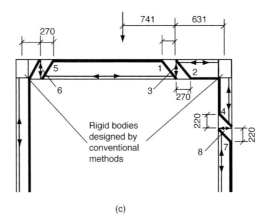

Fig. 4.47. (a) BMD, (b) SFD and (c) physical model of the portal frame in Fig. 4.44 (note: all values correpond to experimental results)

deflection curves and crack patterns at failure respectively, established experimentally for the two frames. Figure 4.48 indicates that the frame designed to the proposed method behaved in a very ductile manner (in fact, it did not suffer any loss of load-carrying capacity throughout the duration of the test which had to be stopped owing to excessive horizontal

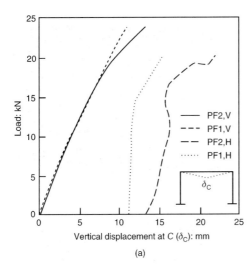
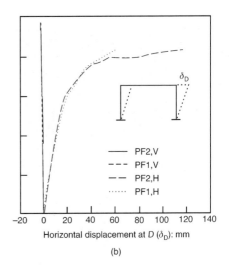

Fig. 4.48. Load versus displacement data for the two frames tested (PFI = code design, PF2 = CFP design): (a) Vertical displacement at point C; (b) horizontal displacement at point D (note that ordinates refer to individual loads)

displacement), in contrast to the frame designed to the code which failed as the maximum load-carrying capacity was attained while exhibiting little, if any, ductility. The cause of failure in the latter case appears to be the excessive cracking of the right-hand side joint (see Fig. 4.49(c)), while, for the case of the former frame, the provision of additional transverse reinforcement at locations IS2 and IS3 led to a more uniform distribution of the cracking throughout the frame which appears to have prevented excessive disruption of the frame joints (see Figs 4.49(e), and compare with Fig. 4.49(a) for PFI).

4.5. The failure of an offshore platform

4.5.1. Background

Despite the gradual acceptance of some of the concepts outlined in the present book (such as, for example, the use of expression (4.1)[4.1, 4.15]) the potential of the CFP method for simple and practical design of structural concrete members under ultimate conditions remains largely untapped. One interesting instance refers to the failure of the structural wall element 'tricell 23' which led to the collapse of the platform 'Sleipner 4' in the North Sea on August 23, 1991, prompting a recent interesting study reported in reference 1.17. It was established that the collapse was due to the failure of the portion of the tricell wall that did not contain stirrups; moreover, the failure of the above portion was found to have occurred not only because the magnitude of the shear force at the wall ends was seriously underestimated by the global finite-element analysis performed, but, also, because the benefits of axial compression on the wall's shear capacity were seriously overestimated by the sectional design procedure used. Furthermore, a comprehensive investigation of the response of RC wall elements in combined axial compression and shear demonstrated that, while ACI 318[4.16] (which under-lied the method used to assess the tricell-wall shear capacity) considerably

DESIGN METHODOLOGY

Fig. 4.49. Cracking pattern just before failure at: (a) B, (b) C, (c) D and (d) E for PF1, and at (e) D and (f) E for PF2

overestimated shear capacity, AASHTO LRFD[4.17] yielded more conservative as well as more accurate shear-capacity predictions.

From results of the experimental investigation reported in reference 1.17, which are reproduced in Figs 4.50 and 4.51, it appears that two additional interesting observations may be made. First, for the particular structural forms investigated, Fig. 4.50 indicates that, in spite of the improved accuracy of the predictions, AASHTO does not always yield a close fit to experimental values (such is the case, for example, of wall PC21), and, secondly, Fig. 4.51 shows that failure of the specimens tested occurred in the middle portion of the walls, i.e.

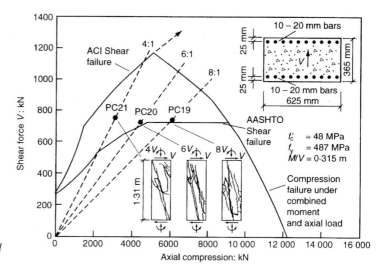

Fig. 4.50 Shear force versus axial compression interaction diagram for the RC walls tested (see reference 1.17)

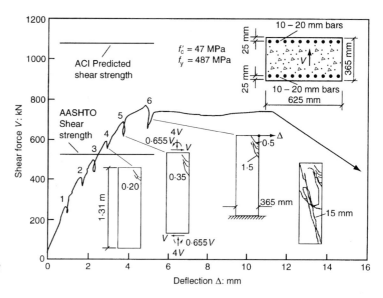

Fig. 4.51 Load versus deformation response of the RC walls tested (see reference 1.17)

in the region of the point of inflection, due to the formation of near-vertical cracking which occurred suddenly and independently from the inclined cracking at the top end of the specimen where current codes predict shear capacity to be exhausted earlier than in other regions of the walls. (It should be noted, however, that the mode of failure may not accurately reflect the causes of failure as it may also be affected by post-failure testing effects.)

In what follows, it is intended to complement the findings of reference 1.17 by demonstrating that the use of the concept of the CFP not only may lead to close predictions of shear capacity, but, also, it may provide a realistic description of the causes of failure. The concept will be used to establish the ultimate limit-state characteristics of both the structural wall elements indicated in Figs 4.50 and 4.51 and the tricell wall component of the structure which appears to have failed under the loading conditions described in Fig. 4.52 (reproduced from reference 1.17), with the ACI and AASHTO predictions being shown in Fig. 4.53 (also reproduced from reference 1.17).

4.5.2 A simple structural evaluation

For the geometric characteristics shown in Figs 4.52 and 4.53, the flexural capacity of the end of the cross-section of the tricell wall (taken as a one-metre wide strip of the (one-way spanning) wall), in the presence of an axial force N satisfying the condition $N/V = 3 \cdot 5$ (where V is the shear force) imposed by the sinking operation (which also yields $M/N = 208$ mm, where M is the moment), can be easily assessed from first principles using the simplified stress block in Fig. 4.13 but allowing also for the presence of an axial force. The flexural capacity (at the fixed-end cross-section) is found to be $M_{fe} \approx 2692$ kNM, corresponding to a pressure of approximately 1688 kN/m² (at a depth of approximately 168 m). These values were used in Fig. 4.54 to

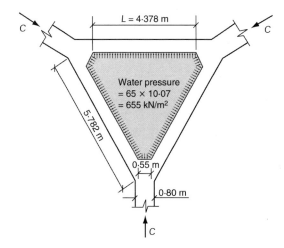

Fig. 4.52 Details of the geometry and loading for the tricell wall 23 of the Sleipner 4 platform (see reference 1.17)

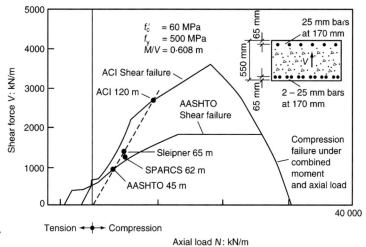

Fig. 4.53 Shear force versus axial load interaction diagram for the tricell wall 23 of the Sleipner 4 platform (see reference 1.17)

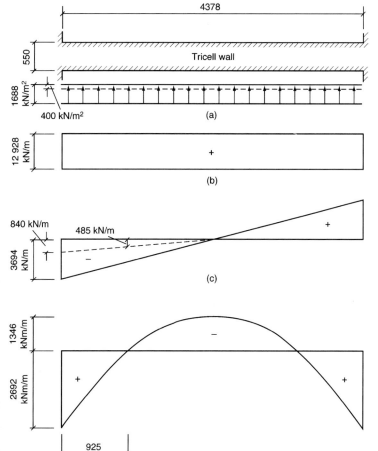

Fig. 4.54 Hydrostatic pressure and internal-force diagrams for the tricell wall: (a) wall and acting pressure (kN/m^2); (b) axial-force diagram (kN/m); (c) shear-force diagram (kN/m); (d) bending-moment diagram (kNm/m) (note: continuous lines refer to flexural capacity, while the dashed lines correspond to some of the maximum predicted values)

construct the internal-action diagrams of which the bending-moment diagram was then used for drawing the physical model of the tricell wall shown in Fig. 4.55. From the latter figure it can be seen that the model comprises two end cantilevers, extending to the theoretical position of the point of contraflexure/inflection situated at a distance of ~ 925 mm from the wall's ends, and a simply-supported beam covering the span between the cantilevers and being supported by them through an 'internal support' that, in the absence of transverse reinforcement, can only be provided by concrete.

Failure of the wall other than flexural may occur at the locations marked 1, 2 and 3 in Fig. 4.55. As discussed in section 4.4.1, the contribution of concrete to the formation of the internal support depends on the tensile strength of concrete which, for design purposes, was recommended to be given a value of 1 MPa. Hence, the total force that can be sustained by concrete in the region of the point of inflection is $f_t b d = 1 \times 1000 \times 485 = 485$ kN which is equivalent to the maximum value of shear force that can develop at the point of inflection. From the shear-force diagram of Fig. 4.54, such a value of shear force at the location of the point of inflection corresponds to a shear force at the end of the wall equal to ~ 840 kN which develops under a pressure of ~ 400 kN/m² (i.e. at a depth of ~ 40 m). (It should be noted, however, that the value of 1 MPa assumed for the tensile strength of concrete is a safe design value adopted within the context of the CFP concept introduced in this book. Under laboratory conditions the tensile strength of concrete may be taken equal to $0.07 f_c$, where f_c is the uniaxial compressive strength of the material.[4.18] For $f_c = 60$ MPa,

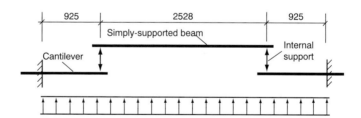

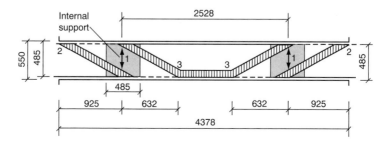

Fig. 4.55. CFP model for the tricell wall

$f_t = 0.07 \times 60 = 4.2$ MPa which cannot be relied upon even in the presence of nominal reinforcement.)

As $a_v/d = 925/485 \approx 1.91 < 2$ (indicating type III behaviour) for the end cantilevers of the physical model of Fig. 4.55, the bending moment M_c that can be sustained at location 2 (fixed-end cross-section) in the presence of both a shear and an axial force can be assessed by interpolating expression (4.1) (but using in the latter s_0 instead of s) as described in section 4.2.2. In fact, equation (4.1) yields $M_c \approx 2100$ kNm ($< M_f = 2692$ kNm) which corresponds to $N_c \approx 10\,000$ kN, $V_c \approx 2857$ kN, and pressure $q_c \approx 1395$ kN/m^2, i.e. failure at location 2 is predicted to occur at a depth of approximately 130 m. On the other hand, for the middle simply-supported beam of the model in Fig. 4.55, for which $L/d = 2528/485 \approx 5.21 < 8$ (indicating type III behaviour), $M_c \approx 2230$ kNm which is larger than the value of 1346 kNm that can be reached at the middle cross-section of the tricell wall at its ultimate limit state (see Fig. 4.54).

It would appear from the above, therefore, that the proposed method predicts that failure of the tricell wall was likely to have occurred at a depth as low as 40 m due to failure of concrete (near-horizontal splitting parallel to the wall faces) in the region of the point of inflection.

4.5.3 Strength evaluation of test specimens

Figure 4.56 shows the design model of the wall elements which were tested for purposes of investigating the causes of the platform collapse, with the load paths used for the tests shown in Fig. 4.50.[1.17] In compliance with the presently proposed design method, the load-carrying capacity of the wall elements subjected to combined axial compression (N) and shear force (V) will be the smaller of the values corresponding to the load-carrying capacities of (a) the elements comprising the walls and (b) their connection (see Fig. 4.56). For $N/V > 2$, calculations show that the load-carrying capacity is dictated by the strength of the connection ('internal support' of the model in Fig. 4.56). Assuming a value of the transverse tensile strength of concrete $f_t = 0.07 f_c = 0.07 \times 48 \approx 3.36$ MPa[4.18] (i.e. the uniaxial value which is essentially independent of the axial compressive stress σ_c perpendicular to it for values of the latter up to about $0.8 f_c (= 0.8 \times 48 \approx 39$ MPa), beyond which it progressively reduces to zero attained for $\sigma_c = f_c$,[1.10, 4.18] the load-carrying capacity is that indicated in Fig. 4.57 which is a reproduction of Fig. 4.50 but now including the values calculated in the present contribution. (The value $f_t = 0.07 f_c$, instead of the value of 1 MPa recommended for design purposes, was selected in order to assess the maximum, rather than a design, force that can be sustained by concrete in the region of the point of inflection. Such a value is considered to provide a realistic indication of the tensile strength of concrete under laboratory conditions.) Note

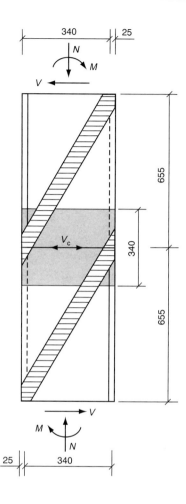

Fig. 4.56. CFP model for the specimens described in reference 1.17

that the strength of the 'connection' remains constant unless the compressive force becomes so large (i.e. $\sigma_c > 0 \cdot 8 f_c$ that it affects the tensile strength in the orthogonal direction: hence, the strength of the 'connection' progressively reduces to zero as N is increased. For $N/V < 2$, the values of the load-carrying capacity of the wall, also indicated in Fig. 4.57, correspond to the shear capacity of the 'cantilever element', the latter being established by using the failure criterion described in section 4.2.2. For $N/V > 2$ (shown as 4:1, 6:1, etc), Fig. 4.57 indicates that the use of the above failure criterion leads to predictions which lie between the values of the shear capacity of the connection and those of the wall load-carrying capacity corresponding to failure due to combined axial compression and bending moment, the latter values of load-carrying capacity also being included in the figure.

Comparing the predicted with the experimetal values, it is interesting to note in the figure that not only does the predicted strength envelope fit very closely the experimental values, but,

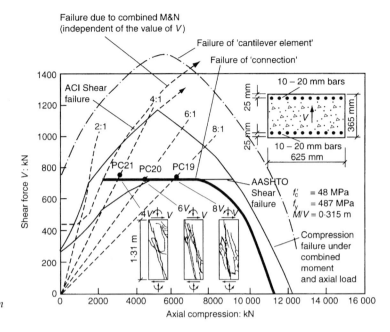

Fig. 4.57. Strength predictions for the specimens described in reference 1.17

also, that the CFP method predicts realistically both the location (near the point of inflection) and the mode (near-vertical cracking originating at this location) of failure. It is also interesting to note in the figure that for $N/V < 1$, say, the values of shear capacity predicted by the CFP method are significantly larger than those predicted by either ACI or AASHTO, indicating a higher contribution of concrete to shear capacity, while for the range $1 \lesssim N/V \lesssim 2$ the CFP estimates lie between the ACI and AASHTO predictions. It should be noted, however, that such a higher contribution of concrete to shear capacity for small values of the axial compression should not be considered to imply that the CFP method necessarily specifies an amount of stirrups smaller than that specified by either of the above codes. This is because, in accordance with the CFP method, the wall behaviour is classified as type III and, for such a type of behaviour, the stirrups specified by the CFP method are intended to increase the bending-moment capacity by an amount $\Delta M = M_f - M_c$ (with M_f being the flexural capacity while M_c is the bending moment corresponding to 'shear' failure), as opposed to the stirrups specified by current codes which are intended to increase shear capacity by $\Delta V = V_f - V_c$ (with V_f being the shear force corresponding to flexural capacity, while V_c is the contribution of concrete to shear capacity) – see section 4.2.5. In the former case the total amount of stirrups required within the shear span is $A_{sv, \text{CFP}} = 2\gamma_s(M_f - M_c)/(af_{yv}) = 2\gamma_s(V_f - V_{c, \text{CFP}})/f_{yv}$ (with a being the shear span), while in the latter case the amount of stirrups required is $A_{sv, \text{CODE}} = \gamma_s(V_f - V_{c, \text{CODE}})/f_{yv}$ (with $V_{c, \text{CFP}}$

DESIGN METHODOLOGY

and $V_{c,\text{CODE}}$ being the values of V_c calculated by the CFP method and the code, respectively, and γ_s the partial safety factor for the reinforcement). It appears from the above, therefore, that although $(V_f - V_{c,\text{CFP}}) < (V_f - V_{c,\text{CODE}})$ (since, as indicated in Fig. 4.57, in general $V_{c,\text{CODE}} < V_{c,\text{CFP}}$), the factor of 2 in the expression of $A_{sv,\text{CFP}}$ results in $A_{sv,\text{CFP}} > A_{sv,\text{CODE}}$ for the case of structural members characterised by type III behaviour.

It is also important to note in Fig. 4.57 that, for walls PC19 and PC20, the CFP method and AASHTO yield similar predictions of shear capacity (i.e. $V_{c,\text{CFP}} = V_{c,\text{CODE}}$). However, while AASHTO specifies stirrups throughout the structural member for the wall to attain its flexural capacity, the stirrups assessed for the same purpose by using the CFP method are required only in the region of the point of inflection with the remainder of the wall being provided with only a nominal amount of transverse reinforcement.

4.5.4 Concluding remarks

THE CFP method yields predictions which provide not only a close fit to the experimental values reported in reference 1.17, but, also, a realistic explanation of the causes which led to the experimentally established mode of failure of the wall specimens. Outside the range of the loading histories which were investigated in reference 1.17, the predictions of the CFP concept deviate significantly from the predictions of either of the codes presently considered. Although the predictions of the CFP method and AASHTO are similar for the shear capacity of the walls PC19 and PC20, the amount of the stirrups specified for the walls to attain their flexural capacity is significantly different, with AASHTO specifying stirrups throughout the structural member while the stirrups assessed by using the CFP method are placed in the region of the point of inflection with the remainder of the wall requiring nominal stirrups.

4.6. References

4.1. The Institution of Structural Engineers and the Concrete Society. *Design and detailing of concrete structures for fire resistance.* (Interim guidance by a joint committee of the Institution of Structural Engineers and the Concrete Society.) The Institution of Structural Engineers, London, 1978.

4.2. Kotsovos M.D. (Kong F.K. (ed.)). Strength and behaviour of deep beams. *Reinforced concrete deep beams.* Blackie, 1990, 21–59.

4.3. Rawdon de Paive H.A. and Siess C.P. Strength and behaviour of deep beams in shear. *Journal of the Structural Division, Proc. ASCE*, 1965, **91**, No. ST5, October, 19–41.

4.4. Smith K.N. and Vantsiotis A.S. Shear strength of deep beams. *ACI Journal*, 1982, **79**, No. 3, May–June, 201–213.

4.5. Ramakrishna V. and Ananthanarayana Y. Ultimate strength of deep beams in shear. *ACI Journal*, 1968, **65**, No. 2, February, 87–98.

4.6. Kong F.K., Robins P.J. and Cole D.F. Web reinforcement effects in deep beams. *ACI Journal*, 1970, **67**, No. 12, December, 1010–1017.

4.7. Rogowski D.M., MacGregor J.G. and Ong S.Y. Tests of reinforced concrete deep beams. *ACI Journal*, 1986, **83**, No. 4, July–August, 614–623.

4.8. Seraj S.M., Kotsovos M.D. and Pavlović M.N. Compressive-force path and behaviour of prestressed concrete beams. *Materials & Structures, RILEM*, 1993, **26**, No. 156, March, 74–89.

4.9. Leonhardt F. *Prestressed concrete design and construction.* (Translated by C.V. Amerongen.) Wilhelm Ernst & Sohn, Berlin, 1964, 2nd edn.

4.10. Lin T.Y. and Burns N.H. *Design of prestressed concrete structures.* John Wiley & Sons, New York, 1981, 3rd edn.

4.11. Abeles P.W. and Bardhan-Roy B.K. *Prestressed concrete designer's handbook.* Viewpoint Publications, Slough, 1981, 3rd edn.

4.12. Kotsovos M.D. and Lefas I.D. Earthquake resistant shear design of reinforced concrete structural walls. *Proc. Int. Conf. on Earthquake Resistant Construction and Design*, Berlin, June 1989, **4**, 569–578.

4.13. Salek S.M., Kotsovos M.D. and Pavlović M.N. Application of the compressive-force path concept in the design of reinforced concrete indeterminate structures: a pilot study. *Structural Engineering & Mechanics*, 1995, **3**, No. 5, September, 475–495.

4.14. Kotsovos M.D. and Michelis P. Behaviour of structural concrete elements designed to the concept of the compressive force path. *ACI Structural Journal*, 1996, **93**, No. 4, July–August, 428–437.

4.15. Council on Tall Buildings and Urban Habitat. *Cast-in-place concrete in tall building design and construction*, McGraw-Hill, New York, 1992.

4.16. ACI Committee 318. *Building code requirements for structural concrete (ACI 318–95) and Commentary ACI 318R–95*, ACI, Detroit, 1995.

4.17. American Association of State Highway and Transportation Officials. *AASHTO LRFD design specifications and commentary*, AASHTO, Washington, 1994.

4.18. Neville, A.M. *Properties of concrete.* Sir Isaac Pitman & Sons, London, 1973.

Coláiste na hOllscoile Gaillimh

3 1111 30090 2390